Residential Care Transformed

Residential Care Transformed

Revisiting 'The Last Refuge'

Julia Johnson
The Open University, UK

Sheena Rolph
The Open University, UK

Randall Smith
University of Bristol, UK

palgrave
macmillan

First published 2010 by
PALGRAVE MACMILLAN

Palgrave Macmillan in the UK is an imprint of Macmillan Publishers Limited, registered in England, company number 785998, of Houndmills, Basingstoke, Hampshire RG21 6XS.

Palgrave Macmillan in the US is a division of St Martin's Press LLC, 175 Fifth Avenue, New York, NY 10010.

Palgrave Macmillan is the global academic imprint of the above companies and has companies and representatives throughout the world.

Palgrave® and Macmillan® are registered trademarks in the United States, the United Kingdom, Europe and other countries.

ISBN 978–0–230–20242–9   hardback

This book is printed on paper suitable for recycling and made from fully managed and sustained forest sources. Logging, pulping and manufacturing processes are expected to conform to the environmental regulations of the country of origin.

A catalogue record for this book is available from the British Library.

Library of Congress Cataloging-in-Publication Data

Johnson, Julia, 1946–
    Residential care transformed: Revisiting 'The last refuge' / Julia Johnson, Sheena Rolph, and Randall Smith.
        p. cm.
    ISBN 978–0–230–20242–9 (hardback)
    1. Old age homes—Great Britain—History.   2. Townsend, Peter, 1928–2009.
I. Rolph, Sheena.   II. Smith, Randall, 1936–   III. Title.
        HV1481.G72J65 2010
        362.610941—dc22

                                                            2010004783

10   9   8   7   6   5   4   3   2   1
19 18 17 16 15 14 13 12 11 10

Printed and bound in Great Britain by
CPI Antony Rowe, Chippenham and Eastbourne

To our parents

Netta Huxley (1902–2002) and David Huxley (1914–84)

Muriel Bell (1915–94) and Thomas Bell (1912–93)

Hilda Smith (1905–81) and Frank Smith (1901–80)

Contents

List of Figures and Photographs

Figures

Photographs

List of Tables

In Memoriam

Peter had agreed to write a Foreword to this book once we had completed our final draft. Very sadly, he died before the text was ready to hand to him. A host of tributes have been published, acknowledging his integrity, his humanity and his seminal writings. This note comprises our reflections of the willing enthusiasm he displayed for our project and the unstinting practical help that he provided. One of the most exciting features of the project was delving into the 'Townsend collection' boxes held in the UK Data Archive at the University of Essex. At times, the material on the homes he and his research team visited presented us with a puzzle. Peter was always happy to help us out. Sometimes, we called upon his memory to test uncertainties in our own minds. At other times we needed practical guidance. For instance, the detail was sometimes lacking about when his research team visited some of the 173 homes in the study, or even where the homes were specifically located. We listed our queries and back came, in that neat handwriting so typical of Peter's meticulousness, the relevant details copied from his personal diaries for the period 1958–9.

Apart from this support, Peter gave willingly of his time to respond to a host of topics we wanted to discuss with him. As the research began to produce newsletters, conference papers, book chapters and journal articles, he again was enormously helpful in commenting on early drafts. In early June 2009, we sent him a draft of one of the Appendixes to this book. It acknowledged and celebrated the invaluable work of our volunteer researchers up and down the country, who traced the fate of the 134 homes which had closed in the years following Peter's study. Peter's reply, as ever, was generous in his encouragement and imaginative in suggesting improvements. His email was sent late in the morning of Sunday, 7 June 2009. That evening he suffered a fatal heart attack. Many thousands of people have benefited – directly or indirectly – from his commitment to the promotion of social justice and the eradication of poverty. We want to thank him and remember him as a generous colleague who gave time willingly to support a project that could not have been undertaken had it not been for his deep commitment to rigorous research and the preservation of research material.

Thank you, Peter.
RS, JJ and SR

Acknowledgements

This book is based on research that was funded by the Economic and Social Research Council (RES-000-23-0995). We are grateful to colleagues who read the proposal before it was submitted, to the anonymous referees who endorsed it and to Robin Means, Sheila Peace, Ian White and Annie Stevenson who, as members of our advisory group, wisely guided us in the conduct of our research. The research involved reusing data from the Peter Townsend collection which is stored in the UKDA. Louise Corti, Libby Bishop and John Southall at the UKDA have shown a consistent interest in our project and Nigel Cochrane, librarian at the Albert Sloman Library at the University of Essex, has been tremendously helpful in searching out information for us. Thanks also to Julie Charlesworth, one of the original archivists of the *Last Refuge* boxes, for the time she gave us and to Robert Pinker who allowed us to interview him about his work as a research officer on Townsend's project.

Several of the photographs reproduced in this book, including the one on the front cover, were taken by Peter Townsend in the late 1950s. We would like to thank Jean Corston, who now holds the copyright, for generously allowing us to use them. The other photographs used in this book are also reproduced with the kind permission of the copyright holders who, in addition to ourselves, include Barbara Prynn and Peter Wade. We also thank Sheila Bennett, Eileen Hayden, Ann James, Keith McCoy, Tony Paine, Philip Roberts and Rita Young for giving us permission to include photographs of themselves or their relatives.

An enormous number of people helped us trace the history of the 173 residential homes visited by Peter Townsend in the late 1950s. We recruited nearly a hundred volunteer researchers across England and Wales. Without their enthusiasm, curiosity and persistence, this project would not have been possible. Each is named in Appendix 1. In addition, we have extended our thanks to numerous local studies librarians and others who helped us to put together some of the missing pieces. Our success in recruiting so many volunteers was in no small part due to the support and efforts of Keith Richards, Lin Jonas and Daphne Sirett, University of the Third Age, Joan Rapaport, History of Social Work Network, Members of OPRSI (Older People Researching Social Issues), Peter Leather, Department of Continuing Education at the University of Birmingham, and various local history associations. Particular thanks to William Evans, a local historian, who piloted the tracing study on two homes.

We also wish to thank the providers, residents and staff of the surviving homes we researched. In particular, we are grateful for the time given to us by the 20 home managers and 75 residents we interviewed together with the 18 members of staff and residents who kept a diary for us. We hope we have treated their words and deeds fairly. Those who wished their contribution to our research to be acknowledged are also named in Appendix 1. We received help too from staff in various local CSCI offices, particularly Lyn Davis and Sandra Gibson from the Avon Office, who we met in 2005.

We thank too our respective universities for the support they have afforded us. June Ayres, Kate Fawcett, Penny Wilkinson, Lesley Henderson, Sandra Riekie, Linda Price and Angela Torrington have provided invaluable administrative and secretarial assistance. Linda Johns and Tanya Hames created and maintained our website and Neil Walker, Vicki Fowler and Kate Stothard provided IT assistance. Several academic colleagues and friends from the Centre for Ageing and Biographical Studies at The Open University and the Research Programme on Ageing and the Lifecourse at the School for Policy Studies, Bristol University, have helped us throughout the research and we are particularly grateful to Chris Harris, Ken Blakemore and Moyra Sidell who read earlier drafts of the manuscript. We are deeply indebted to Henry Rolph for his interest and support in the conduct of the research and very special thanks to Bill Bytheway who has spent many hours patiently discussing, advising and helping us with the research and the preparation of this book.

Finally, we would like to thank Olivia Middleton and Philippa Grand, our commissioning editors from Palgrave Macmillan, and Manavalan BhuvanaRaj, our project manager, for their patience and support in the production of this book. Some of our chapters draw on papers we have had published elsewhere. Part of Chapter 4 draws on Johnson et al. (2010), published by Cambridge University Press, and parts of Chapters 3 and 9 draw on Johnson et al. (2007) and Rolph et al. (2009), published by The Policy Press and Taylor & Francis respectively.

List of Abbreviations

BASW	British Association of Social Workers
CQC	Care Quality Commission
CRB	Criminal Record Bureau
CSCI	Commission for Social Care Inspection
CSIW	Care Standards Inspectorate for Wales
CSSIW	Care and Social Services Inspectorate for Wales
DH	Department of Health
DSS	Department of Social Security
ESRC	Economic and Social Research Council
HI	Home Interviews (files held at the UKDA)
HIMP	Health Improvement Programme
HR	Home Report (written in 2005–6)
LRR	Last Refuge Revisited
LSE	London School of Economics
MI	Manager Interview (conducted in 2005–6)
NCSC	National Care Standards Commission
NHS	National Health Service
NSF	National Service Framework
NVQ	National Vocational Qualifications
OPRSI	Older People Researching Social Issues
PAI	Public Assistance Institution
POVA	Protection of Vulnerable Adults
PTC	Peter Townsend Collection (held at the UKDA)
RD	Resident Diary (written in 2005–6)
RI	Resident Interview (conducted in 2005–6)
SEAN	State Enrolled Auxiliary Nurses
SRN	State Registered Nurses
SSI	Social Services Inspectorate

SWHN	Social Work History Network
U3A	University of the Third Age
UKDA	UK Data Archive (University of Essex)
WAG	Welsh Assembly Government

Part I The Context

1
Why Revisit *The Last Refuge*?

In January 2004, two of us struggled across the windswept campus of the University of Essex. We were heading for the library where the Peter Townsend collection is deposited. A room had been set aside and the two boxes we had asked to see were sitting on the table waiting for us. It is hard to describe the excitement we felt. For the first time we were to see and handle the old typewritten reports, the rusting paper clips and staples, and scribbled notes in what was clearly the handwriting of Peter Townsend. Here were the original data from which, nearly 50 years previously, he had produced his classic book *The Last Refuge* (Townsend, 1962) which reported his study of residential care for older people carried out in the late 1950s.

Our book, this book, tells the story of his research and our research, and it highlights the importance of a longitudinal perspective in making sense of the present. It is a book about continuity and change over the last 50 years not just in residential care for older people but also in the way old age is researched, represented and theorized.

A work of lasting importance

The Last Refuge is a very substantial piece of work; the unabridged version of the book is 552 pages which include 113 pages of detailed appendixes and 38 photographs, most of which were taken by Townsend himself. It reports a national study of residential care for old people provided in England and Wales under the National Assistance Act 1948[1] and is a devastating critique of residential care for older people in the late 1950s. It became a hugely influential book which has been read by generations of academics, students, social workers and policymakers.

As an anonymous reviewer of the book commented:

> Peter Townsend has done for old people what the Curtis Committee did for deprived children. One can only hope that his work will bring about a comparable revolution in official attitudes and policies. This book is on the scale of a Royal Commission Report – one of the legendary nineteenth-century Royal Commissions to which modern social critics are always regretfully looking back. It is an astonishing achievement for one man (and despite the generous tribute he pays to those who helped him with the project, this is very much the work of an individual).
>
> (Anonymous, 1963, p. 150)[2]

One reason why this book has become legendary and has been so influential is that it is not just an extremely rigorous scientific text but also a literary text. In this sense the book is very much 'the work of an individual'. Geertz has drawn attention to the anthropologist as author and has observed that 'it might be difficult to defend the view that ethnographic texts convince ... through the sheer power of their factual substantiality' (1988, p. 3). Townsend is a gifted writer and one cannot avoid being moved by his prose. This, combined with his photographs, creates a powerful, memorable and convincing book.

The research for *The Last Refuge* was conducted in the late 1950s when Townsend, who was already on the path to becoming an extremely eminent author and scholar,[3] was a newly appointed lecturer at the London School of Economics (LSE). Here he worked under Richard Titmuss, the first professor of social administration in Britain. He first met Titmuss when he was a research assistant at Political and Economic Planning (the precursor of the Policy Studies Institute). Before moving to LSE, he was working at the Institute of Community Studies set up in 1954 by Michael Young in Bethnal Green, East London. As Evans and Thane (2006) point out, the institute 'encouraged descriptive, empirical, empathetical study of working class life, with the explicit policy aims of describing social conditions in order to advocate policy change'. Townsend conducted his first major piece of research when working there – a community-based study of the lives of older people in the East End of London – which resulted in his first classic publication, *The Family Life of Old People* (Townsend, 1957). Townsend was drawn to LSE by his admiration for Titmuss who he considered to be 'just the kind of person with the kind of values I believed in' (Page, 2002, p. 2) and from whom he felt he could learn. Consequently, when in 1957 he was successful in acquiring a grant from the Nuffield Foundation to undertake a study of institutional care for old people, he took up a position at the LSE.

The origins of Townsend's research for *The Last Refuge* lay in a single visit he made in June 1955 to a very large former Public Assistance Institution (ex-PAI)[4] in London. He was both daunted and shocked by what he found (Townsend, 1962, pp. 4–6; Thompson and Townsend, 2004): bare, overcrowded dormitories with iron-framed beds set close together; 'bleak and uninviting' dayrooms where 40 men in shapeless clothes and slippers sat side by side without speaking, inactive, staring straight ahead as if life had been 'drained from them, all but the dregs'. Townsend highlighted their loneliness in the midst of others, describing isolated figures sitting alone in the washrooms, standing still in a corridor, and one man 'looking out of a staircase window weeping silently'. Little privacy or dignity remained, and there was evidence of an insensitive and regulatory regime, with some staff regretting the passing of the Poor Law, and with it their power to control and punish. This visit led him to question how such a situation could still exist some ten years after the founding of the welfare state. Although it had been the aim of the post-war government to close the ex-PAIs and to accommodate older people in smaller converted or purpose-built homes, this had not happened. Indeed Townsend was surprised to discover subsequently that just over three-fifths of people living in residential care homes were still being accommodated in ex-PAIs. Townsend realized that the plight of older people hidden away in institutions and other residential care homes required urgent attention. Hence, with its focus on institutions, *The Last Refuge* complemented his earlier study of older people living in the community.

In an interview with Robert Page in 2002, Townsend revealed that he started his research for *The Last Refuge* with a definite hypothesis. In recalling a conversation with Brian Abel-Smith at the time, he says:

> I was chewing over how we would handle the huge quantity of information. Brian in his usual gung-ho way said, 'Come on Peter, what is your conclusion? What do you think you're going to say?' I said, 'I suppose I'm trying to work with the idea of whether residential homes for old people are the right solution or not?' He said, 'That's it. You've got to take a view. You think they shouldn't exist don't you?' 'Well, I don't know. It's for me to find out but maybe I should take that as a hypothesis'.
>
> (Page, 2002, pp. 3–4)

Photograph 1.1   Peter Townsend (left) with Brian Abel Smith (right) at the London School of Economics in the late 1950s
Source: Peter Townsend. © Jean Corston

His fundamental research question was, therefore: 'Are long-stay institutions for old people necessary in our society and, if so, what form should they take?' (Townsend, 1962, p. 3).

Townsend studied social anthropology at Cambridge University. At that time the dominant paradigm was structural functionalism and this has had a lasting impact on how he has approached his research. As he says,

> the functionalist theme of social anthropology was that each piece of behaviour, each system of relationships, was functional in relation to the rest of society. ... Although the idea that societies are 'systemic', is something we find so overwhelming, that we often depart from it, and we retreat into small-scale studies.
>
> I'd still take that view ... about generalism and specialism. ... You can't be a good specialist unless you're a generalist also, and you can't be a good generalist unless you have some specialist interest.
>
> (Thompson and Townsend, 2004, p. 88)

Central to his design, therefore, was the combination of both quantitative and qualitative methods. He conducted a national survey of all 146 local authorities in England and Wales in order to document the extent and nature of residential care provision for older people. This included interviewing 65 chief welfare officers.[5] He then drew a stratified random sample of 180 local authority, voluntary and private residential care homes, 173 of which were visited by him or members of his research team. He had learnt the importance of sampling when working on family life in the East End of London and of how a properly drawn random sample incorporates the extremes of experience (Thompson and Townsend, 2004). In the case of *The Last Refuge* the extremes he captured ranged, for example, from very large ex-PAIs, located in large cities and accommodating more than 1000 people, to small privately owned homes by the seaside accommodating six or less residents; from homes costing three-and-a-half guineas a week to those costing 30 guineas (equivalent at 2009 prices to between about £60 and £520).[6] At the same time, Townsend was committed to the anthropological method whereby you immersed yourself in the society you were studying through participant observation. He realized the potential of this method for studying our own society and, while undertaking research for *The Last Refuge* he spent some time working as a bathing attendant in one of the institutions he studied.

He and his research officer, Robert Pinker,[7] conducted 80 per cent of the fieldwork. They visited the large institutions together or with other members of the research team. In 2005, Townsend transcribed for us his 1959 appointments diary. This provides a fascinating insight into the punishing schedules these two young men set themselves. By way of example, between 7 and 30 April 1959, Townsend alone visited homes and interviewed chief welfare officers in East Sussex and Brighton, Surrey, London, Chichester, Trowbridge, Stoke-on-Trent, Buxton, Lancashire, Liverpool and Cumberland. The schedule for adjacent months was equally busy and diverse. In an interview with Robert Pinker in 2007, he described to us how they 'did the country' speeding around in 'a funny old Ford van' talking endlessly. And it is worth bearing in mind that, with the exception of the Preston bypass, there were no motorways at this time.[8] During the course of each visit they interviewed the warden or matron of the home, toured around the home noting its furnishings and other facilities and interviewed a sample of newly admitted residents. Influenced by the *Picture Post* genre of photography, Townsend also used his Leica camera to take nearly 100 photographs in the homes he visited (Townsend, 2005). Pinker also described how on joint visits, while one was interviewing, the other would look around. Sometimes they came across 'some pretty grim side wards'.

Photograph 1.2 Robert Pinker (right) with the matron and warden when visiting
The Shrubbery, Walsall, a local authority home which closed in 1988
Source: The Peter Townsend Collection. © Jean Corston

In interviewing residents, Townsend was particularly concerned to
assess whether they needed residential care and, to this end, he devised
a new research tool to measure the 'incapacity for self-care' of residents
(1962, Appendix 2). For comparative purposes, he also devised a tool
for measuring the quality of homes (1962, Appendix 3). Following the
visit to each home, a report was written and these reports, together with
copies of the photographs, are now deposited in the UK Data Archive
(UKDA) at the University of Essex.

In keeping with the style of ethnographic reporting at the time, the
home reports contain vivid descriptions of the homes visited and the
people interviewed. From today's point of view, some of these may seem
surprisingly candid, if not judgemental. For example, one proprietor is
described as 'an ungalloping major ... he assumes all the mannerisms
and postures of the regular officer retired from a very good regiment.
He had a mildly egg-shaped head and a moustache.' Another proprietor looks 'a little bit like Mervyn Johns' and an assistant matron 'rather
like Margaret Rutherford'. The matron of a voluntary home whose
residents are 'mostly drawn from the middle class' is described as

'a tall spindly woman of 42. ... Her face is rather pock-marked and she seemed to be of somewhat lower social status than many of the old people in the home.' And the warden of a local authority home has 'swept back grey white hair, a paunch with two or three fly buttons undone, a very marked cast in one eye, a thick nose and a fleshy face'. This kind of detailed description was invaluable in providing a picture of the homes and the people who staffed them. As Richard Crossman (1962, p. 930) who reviewed the book comments, Townsend managed to balance these 'impressionistic descriptions of what he and his colleagues saw for themselves, with unassailably objective conclusions reached by the statistical analysis of information derived from carefully composed questionnaires'. Thus, Crossman concludes, he 'successfully combines the moral passion of the engaged observer with the scientific attachment of academic analysis' (ibid.).

Although Townsend judged a majority of homes visited to be of poor quality, particularly the local authority homes and the larger homes, some he found were providing a particularly high standard of care, most notably those provided by the voluntary sector. Hence his research uncovered gross inequalities in the quality and quantity of provision for people with the same needs. Townsend also found that about half of all residents were not physically or mentally incapacitated and that people were being admitted to residential care as much for social as for physical reasons. For example, 47 per cent of the residents of the ex-PAIs were men, many of whom were living in these institutions simply because they had nowhere else to live. These findings led him to question the assumptions that people chose to be in residential care and that they wanted to stay there permanently. He concluded that residential care as an instrument of social policy for older people should be abandoned. Those who were seriously incapacitated should be in hospital or small publicly owned nursing homes. All others, he argued, could live in private households provided their housing was appropriate and the necessary domiciliary support was available.

The impact of Townsend's research

Townsend's book had a substantial impact not just on policy and practice in the UK but also on research methodology and the development of social theory in relation to ageing. Like Crossman, the reviewer for *The Times Literary Supplement* quoted above noted that the book was 'the first really satisfactory attempt to reconcile the participant observer techniques of anthropology with modern statistical methods'

(Anonymous, 1963, p. 150). In addition, as we discuss further in Chapter 2, Townsend's research tools were developed by subsequent researchers in the field of residential care to measure dependency and quality of life.

Townsend's findings also influenced the way in which he subsequently theorized the position of older people in Britain. In a seminal paper (Townsend, 1981) he argued that the dependency of 'the elderly' was socially created through, among other mechanisms, the unnecessary placement of people in residential care homes where they lapsed into apathy and despair. His theory of structured dependency contributed to the early development of the political economy of old age, a body of theory that has been highly influential in gerontology in the UK and beyond (see, for example, Estes, 1979; Guillemard, 1980; Walker, 1981; Phillipson, 1982; Myles, 1984; Quadagno, 1988; Minkler and Estes, 1991; Minkler and Cole, 1991) and that has continued to be developed (Bengtson and Schaie, 1999; Estes et al., 2003).

However, it has been argued that the impact of Townsend's research on policy and practice has been negative as well as positive. Some, like Kathleen Jones (1962), for example, were critical of his recommendations. Although she was full of praise for the book, she thought it 'marred by some very subjective judgements' and argued that

> [o]ld people, like young people, can be highly ambivalent, and their actual field of choice (as opposed to their ideal field of choice) can be very restricted. ... This is not really a book about old people, for their attitudes, emotions and reactions are judged entirely by middle class, young middle aged standards. It is a book about the author's belief that institutional living is undesirable.
>
> (Jones, 1962, p. 31)

The Last Refuge did indeed become part of a stream of anti-institutional literature published in the 1960s and 1970s, sometimes referred to as 'the literature of dysfunction' which was highly critical of institutional care (Jones, 1967; Jack, 1998). This literature comprised a number of sociological studies of institutional life (Barton, 1959; Goffman, 1961; Henry, 1963; Robb, 1967; Morris, 1969; Meacher, 1972; Miller and Gwynn, 1972). This, together with a series of official inquiry reports following serious allegations about the prevailing conditions in long-term hospitals, such as Ely (Howe, 1969), Farleigh (Watkins, 1971), Whittingham (Payne, 1972), South Ockenden (Inskip, 1974) and Normansfield (Sherrard, 1978), formed a powerful critique which has had a lasting impact on attitudes towards institutional care. Jack has argued that the effect of it has been 'to relegate residential institutions to the status of last resort' (1998, p. 1). It was a literature therefore, as we discuss further in Chapter 2, which made a powerful contribution to promoting the move towards community care policies. Townsend's recommendations about the replacement of residential care by appropriate housing, for example, led to an enhanced interest in sheltered housing and to a major development of this alternative in the following decades (Ministry of Housing and Local Government, 1962a, 1962b; Butler et al., 1983). Likewise, his recommendations speeded up the demolition of the old PAIs and the development of a new building programme for smaller, purpose-built residential homes for older people (Peace et al., 1997).

Despite these developments and Townsend's recommendations, in both England and Wales, alongside the increase in the number and proportion of older people in the population, there has been a steady increase in the number of places in residential care since the late 1950s. Indeed, residential care has continued to be a major plank of provision and expenditure for older people. In 2005–6, when our own research was conducted, councils in England were spending nearly twice as much on residential care placements (and three times as much if expenditure on nursing home placements is added in) as on home care for older people (CSCI, 2008, p. 161). At the same time, concern about the future (and funding) of long-term care remained a dominant theme in policy debates. As we discuss in Chapter 2, by the 2000s, the neoliberal choice agenda was being linked to ideas of user empowerment and control, leading to what has been labelled the 'personalisation' of social care (DH, 2008). It is striking that the place of residential care is rather muffled in this context. The emphasis is on care in the community (a slippery phrase) and the profile of residential care is as a negative resource of last resort. Yet substantial public resources are tied up in what has become a predominantly private sector operation.

Given the powerful impact of *The Last Refuge* on research, policy and practice, we were curious to know what became of the homes Townsend visited and of which he was so critical. What could be learnt about the history of residential care by returning to his work and addressing this question?

Revisiting *The Last Refuge*

We first discussed the idea of revisiting *The Last Refuge* in November 2003. The deposit of data from Townsend's lifetime of work in the National Social Policy and Social Change Archive at the University of

Essex (now the UK Data Archive) provided us with the opportunity to revisit his study of residential care. As Thompson has commented, the Peter Townsend collection is 'very likely the most in-depth documentation that will ever be collected of the conditions and experience of old age and poverty in Britain, a unique and unrepeatable set of qualitative research material' (Corti and Thompson, 2004, p. 338).

Although there are few contemporary studies of residential care which are as extensive as Townsend's, there have been a substantial number of studies which provide cross-sectional data. Our intention was not, therefore, to replicate Townsend's original study by drawing a new, representative random sample of homes so as to compare the national picture then and now. Rather, we wanted to draw on his data to develop a longitudinal study or what Wadsworth (2002) has described as an 'accelerated prospective' or 'catch-up' study. By focusing on a single cohort of homes, we could explore the historical processes of continuity and change at the level of individual homes. As Davies has observed: 'The principal strengths of longitudinal studies of all sorts lie in their greater sensitivity to change, the increased likelihood of being able to distinguish fluctuations from fundamental changes, and the greater depth of ethnographic understanding achieved from the multiple perspectives that such research facilitates' (1999, p. 175). Our plan was to find out what happened to the 173 homes Townsend visited and to revisit a sample of the surviving homes. Although, the conclusions that can be drawn on the basis of a population of surviving homes may be limited in some ways, there is, as Wadsworth (2002) points out, something to be learnt from the non-survivors as well.

The principal aim of our first visit to the archive in January 2004 was to find out if it was possible to acquire sufficient information on the 173 homes Townsend visited in the late 1950s in order to trace their subsequent history. We managed to assemble a list of the homes, sometimes with barely any information on precise location. We were then able to start finding out which, if any, had survived and were functioning as care homes. When we were confident that at least 20 were still in existence, we began drawing up a proposal to put to the Economic and Social Research Council (ESRC). Our aim, however, was to trace the history of all 173 homes. This would enable us not only to follow up and revisit the survivors but also to find out what became of the homes that did not survive, when (and if possible why) they ceased to function as care homes. The only way to do this, we decided, was to recruit locally based volunteer researchers. Before submitting our proposal, therefore, we had to test the viability of this strategy. First we had to find out precisely what would be involved in locating and then tracing the history of a home that might have closed at some point during the last 50 years. Secondly, we had to find out if there were older people 'out there' who would be willing to volunteer their services to assist with this. After a substantial amount of preparation, our proposal was finally submitted to the ESRC and accepted for funding. In consequence, Sheena Rolph joined the project when it commenced officially in May 2005.

The current preoccupation of policymakers is how to afford long-term care for vulnerable people. Our research has been concerned with longer-term questions which, nevertheless, have a bearing on the present: What kinds of home are able to continue to function from one decade to another? What events have precipitated closures? As one generation of residents and staff succeed another, what kinds of adjustments and adaptations help to sustain continuity? And why do policymakers and members of the general public remain ambivalent about care home provision? If care homes are to offer older people a real sense of 'home' for their remaining years of life, these kinds of questions, relating as they do to the stability or otherwise of residential care, are extremely important.

As well as contributing to the policy debate on long-term (including residential) care, another aim of our research was to contribute to methodological debates and developments. Ever since Stacey and colleagues (Stacey et al., 1975; Bell, 1977) revisited the Banbury study (Stacey, 1960), there has been a developing interest in studying social change through revisiting classic sociological studies. For example, Phillipson and colleagues (2001) restudied three urban areas which had been the locations of a number of community studies conducted in the 1940s and 1950s (Sheldon, 1948; Townsend, 1957; Young and Willmott, 1957; Willmott and Young, 1960) and Charles and colleagues (Charles et al., 2008) restudied Rosser and Harris' study of family and kinship conducted in the early 1960s (Rosser and Harris, 1965). These restudies, while neither replicating the previous studies nor drawing on unpublished data, used formerly researched communities to investigate social change. Since 1996, the ESRC has required researchers it funds to offer their data for deposit in the UKDA and it has a particular interest in developing the use of archived data (Corti and Thompson, 2004). Consequently, there is now a substantial sociological literature emerging relating to the reuse and secondary analysis of archived qualitative data (Heaton, 2004; Moore, 2007; Macleod and Thomson, 2009) and more social science researchers are drawing on such data. Evans and Thane

(2006), for example, have drawn on Dennis Marsden's original data, upon which his book *Mothers Alone* (Marsden, 1969) is based and which is now archived in the UKDA, to explore the history of single motherhood. Goodwin (2005) using as yet unarchived data has followed up Norbert Elias' unfinished study of young workers conducted in the early 1960s, and Nigel and Jane Fielding (2008) have reanalysed Cohen and Taylor's (1972) original and archived data on long-term imprisonment. These studies raise questions not only about the boundaries between historical and sociological research but also about how the research design should be conceptualized, whether as a restudy, a replication, revisiting, a follow-up, a reanalysis, reuse or qualitative longitudinal research (O'Connor and Goodwin, 2008). We describe our own research as 'revisiting' rather than 'restudying' and discuss our design more fully in Chapter 3.

In summary, this book has a number of aims. First, through our examination of continuity and change in residential care over the last 50 years, it aims to contribute to debates about the future care of older people in the UK in the twenty-first century and the place of residential care in that debate. Secondly, it aims to make a contribution to the discussions surrounding revisiting studies and the reuse of archived data. Thirdly, because we involved nearly a hundred people in the task of tracing the subsequent history of the homes Townsend visited (see Johnson et al., 2006; www.lastrefugerevisited.org.uk), almost all of whom were aged 60 or more, it aims to contribute to our understanding of involving older people in research, a very topical issue in the early 2000s. Finally, in revisiting work carried out 50 years ago we draw some conclusions about the changes and continuities in the imagery of old age and in the way that old age is theorized.

The plan of the book

In Chapter 2 we focus on the broader social, policy and research contexts within which Townsend's and our own research have been conducted. In many ways, society at the beginning of the twenty-first century is barely recognizable when compared with that of 50 years ago. In consequence, the experience of being old in both periods is very different, as are policies for older people and the way that old age is researched, understood and theorized. This chapter examines the shifts in policies for older people in regard to residential care and the relationship between these policies and the development of research in this field both in the UK and beyond. In the early decades of the welfare state, policy and research were dominated by public sector provision and underpinned by a commitment to collective responsibility. By the beginning of the twenty-first century, residential provision for older people has been almost entirely privatized and the policy discourse is very different. The emphasis is on individual choice and responsibility, empowerment, user control and the 'personalisation' of social care services through devices such as direct payments or individual budgets. Despite these changes there are continuities: in both periods and in the years between, the call on the public purse has been an issue of policy concern. Townsend's research, we argue, had an important impact on subsequent research in the field of residential care. His methods of inquiry into the quality of care and the characteristics of residents in care homes have had a lasting impact. In more recent years, with the arrival of the mixed economy, issues of regulation and inspection became more prominent, as did the idea of user engagement in the research process. Parallel to this, research approaches have reflected a move away from a broadly positivist position, where little attention was given to the relationship between the researcher and the researched, to a more reflexive one where the impact of this relationship is recognized.

Chapter 3 outlines in detail the design of our 'revisiting' study. It describes how we traced (the tracing study), through the engagement of nearly a hundred volunteer researchers, what happened to the 173 homes Townsend visited and how we followed up (the follow-up study) those still functioning as care homes some 50 years later. This chapter raises a number of issues in relation to the reuse and secondary analysis of archived data. It also raises issues regarding the engagement of volunteer researchers (predominantly older people) in funded academic research. Additionally, our research raises many ethical issues relating not only to the reuse of archived data but also to current research practices in care homes for older people.

Part II, 'Revisiting *The Last Refuge*', reports on the findings of our research. Chapter 4 focuses on the findings from the tracing study to establish what kinds of home survived and which ones closed. It describes what happened to the buildings of the homes that closed and the sites upon which they stand or stood. It also looks at factors associated with survival and the kinds of strategies adopted by providers in order to stay in the 'market' for residential care. The following four chapters (Chapters 5–8) compare the surviving homes then and now. Drawing on data on staff and residents, both then and now, in 20 of the 37 surviving homes still providing permanent residential care, the focus of Chapter 5 is on those who lived and worked in the homes in the two time periods. Chapters 6 and 7 report on the qualitative data collected in the 20 homes we revisited by touring the buildings, interviewing the manager and a number of recently arrived residents, taking photographs and asking some people to keep a diary for a week. These chapters provide vivid evidence of continuity and change in the daily lives of residents. The final chapter of Part II of the book addresses the quality of care in the homes. Townsend's measures were used to compare 1958–9 and 2005–6. These were then compared with the measures used by the Commission for Social Care Inspection in England (CSCI) in 2005–6. Finally, the process of regulation and inspection then and now was analysed in the light of Townsend's concern about standards.

Part III of the book comprises two chapters. Chapter 9 reflects on the research approach underpinning 'revisiting' studies, particularly qualitative longitudinal research and the reuse of archived data. It picks up on issues raised in Chapter 3 where we discuss our research design. Addressing and resolving methodological and ethical issues were key challenges. The impact of using older people as researchers 'outside the university walls' (Finnegan, 2005) was a further key theme of the inquiry. The final chapter focuses on continuity and change in residential care. Townsend's recommendations and conclusions are compared with the results of this 'revisiting' study. It is hoped that the evidence produced will help in the crucial policy debate on the future funding of long-term care and the transformation of social care provision. What should be the relationship between the individual and the state and what should be the essence of social policies for older people in the UK?

The anonymous review of *The Last Refuge* in the *Times Literary Supplement* concluded that this study deserved to become a classic of social research. 'It will be read long after the conditions which provoked Peter Townsend to undertake his work have passed into the pages of history books to be gasped over by the children of a more humane generation' (Anonymous, 1963, p. 150). Continuity or change? Has society become more humane or not?

2
Changing Contexts of Care

Against the backdrop of two world wars and the depression of the 1930s, growing older in the 1950s had particular meanings and carried particular memories. Older people were experiencing the slow transformation of society through times of rationing, poverty and hardship, into times of optimism, new opportunities and the energetic rebuilding of cities and communities. As has been pointed out, however, 'the lives of those who are growing old today cannot be the same as the lives of those who grew old in the past or those who will grow old in the future' (Riley et al., 1999, p. 333) and 50 years on, the backdrop to growing older is very different. The main focus of this chapter is on the policy and research contexts within which people have aged and our research and Townsend's research were conducted.

As we demonstrate later in this chapter, there have been substantial shifts in the way older people have been represented in both policy and research discourses. Nevertheless behind the changing rhetoric, there have been significant continuities. As Pickard argues, official discourses relating to old age policy and practice have over several historical eras, taken for granted 'both a low priority for older people, and the burden that ageing places on the state and its institutions' (2009, p. 81). A major constant has been the use of 'apocalyptic demography' (Robertson, 1990, 1997) to articulate concerns about the 'problems' or 'challenges' of an ageing population and how they should be addressed.

An ageing population

Katz (1996), in reviewing the history of debates on the age structure of the population, identified links between demographic alarmism and gerontological knowledge:

> Alarmist demography is based on the Malthusian ethos that, if allowed to go unchecked, the population growth of working, dependent, and marginalised classes will inevitably create massive economic and political instability. Alarmist demography and gerontological knowledge came together in the social surveys of the late nineteenth and early twentieth centuries that decried the growth rate and poverty of the elderly population as an economic and moral crisis.
>
> (p. 69)

The speculations of some pre-war commentators reflected this demographic pessimism. In 1939, for example, when 14 per cent of the population of England and Wales were over the age of 60, Leybourne-White and White (1945) predicted that by 2000 they would account for 29 per cent and, by 2019, 35 per cent of the whole population. They expressed deep anxiety about what they saw as the inexorable growth in the proportion of older people in the population:

> [T]he quality of our social life is bound to change because of the growing preponderance of old people. ... No nation can welcome such a development. For social purposes, therefore, the best policy would be to regain as quickly as possible a balanced population, with more young and fewer old people.
>
> (op. cit., p. 43)

After the Second World War, it was assumed that women would return home from wartime jobs to domesticity and child rearing, and great attention would be paid to rejuvenating the population and to the health and wellbeing of children. Indeed, the Royal Commission on Population (1949) argued for family-friendly policies that would boost the birth rate particularly among 'the better educated and more intelligent' (quoted in Means and Smith, 1998, p. 211). This emphasis on building a new nation around the potential of young people had implications for policies relating to older people. As Beveridge, architect of the post-war welfare state had made clear:

> It is dangerous to be in any way lavish to old age, until adequate provision has been assured for all other vital needs, such as the prevention of disease and the adequate nutrition of the young.
>
> (Beveridge, 1942, para. 236)

Thane (2005, p. 9) suggests that old age is invariably presented as 'a story of decline', parading a list of problems associated with helplessness, lack of independence, and healthcare. As she points out, the 'gloomy' population projections 'produced by the then novel techniques of demography' (Thane, 2000, p. 13) were usually incorrect. Indeed, migration from the Caribbean during the 1950s together with the increased number of women in the labour force completely undermined predictions about population decline and what later came to be referred to as the 'dependency ratio'. Commentators such as Titmuss, writing in the 1950s, challenged the view that an ageing society was a problem and claimed that 'the demographic changes which are under way and are forseeable have been exaggerated' (Titmuss, 1958, p. 56). Rather he viewed the decline in infant mortality and the increase in life expectancy for the working classes as an achievement of the twentieth century.

Means and Smith (1998, p. 213) have argued that 'the demographic debate of this period created a climate of concern about "the burden of dependency" caused by elderly people', a concern that focused not just on the cost of pensions but also on health and welfare services, particularly the National Health Service (NHS) and 'bed blocking' which fuelled the growth of residential care. During the economically buoyant 1960s and early 1970s, services for older people were not a high priority politically and the demographic debate subsided (op. cit.). However, following the retrenchment from the late 1970s, the 1980s saw the re-emergence of demographic alarmism (Mullan, 2000). Research programmes to tackle the issues of what was construed as a rapidly expanding and infirm population of elders were set up, with titles such as 'The Rising Tide' (Health Advisory Service, 1982). And predictions about the future preponderance of older people in society have continued to the present day. Comparative statistics as in the 1950s, while appearing to be factual and objective, serve to foster alarm. For example, the opening chapter of the Office for National Statistics *Focus on Older People* report, published in 2005, states:

> In 1901 nearly one person in seven (15 per cent) were aged 50 and over. This had increased to one in three by 2003 and is still rising. By 2031 it is projected that over 40 per cent of the total population will be aged 50 and over. The older population is currently growing twice as fast as the population as a whole.
>
> (Soule et al., 2005, p. 2)

When, in August 2008, it was announced that 'the ageing index', which is the ratio of older people to children, had for the first time passed 100.0, this became a major media news story. The *Daily Mail* dramatically reported:

Pensioners outnumber children for the first time in Britain's history, startling figures revealed yesterday. The astonishing milestone follows years of steadily rising life expectancy and a significant fall in the number of children and young teenagers. Experts described the watershed moment as a 'wake-up call', warning of grave implications for many aspects of national life including the Health Service, social care for the elderly, pensions and housing.

(Hickley and Greenhill, 2008)

Similarly Magnus (2009, p. 28), a senior economic adviser for UBS investment bank, invoked a variety of melodramatic metaphors, in writing in the *The Guardian* newspaper of a 'dependency timebomb', 'the seismic impact of a rapidly ageing population' and 'the baby boomer retirement avalanche'.

The evidence is, however, that older people make a substantial contribution to the welfare of society through reciprocal activities in social networks, as supporters of families (whether large families in the 1950s or single-parent families and families with working mothers in the early twenty-first century) or through continuing in meaningful paid or voluntary work. Furthermore, as Thane comments, older people 'make fewer demands on natural resources, create less waste, and require services such as public transport, small-scale and accessible shopping, and local parks and recreation', all features of the 'sustainable' society to which many aspire (Thane, 2005, p. 492). Despite this, policies for older people both now and 50 years ago have been dominated by what Phillipson et al. refer to as 'the constant alarmism about the so-called burden of an ageing society' (2001, p. 259).

Ironically, it is the post-war baby boom generation that is now becoming the focus of concern and it is the numbers of very old people that are increasing most sharply. The 2001 Census, for example, reported nearly 8600 centenarians, the age group with the highest growth rate in the whole population, 'roughly doubling every ten years between 1951 and 2001' (Soule et al., 2005, p. 4). The percentage of people aged 65 or more who live in care homes is very small. It is of course very old people who, on the whole, occupy them today. In 2001, the average age of care home residents in England was 85 years and one in four of people aged 90 and over in the UK lived in a care home (Help the Aged, 2008). It is difficult to provide absolutely comparable statistics for 1960, but just 42 per cent of the male residents and 54 per cent of the female residents in Townsend's representative sample of homes were aged 80 or more (Townsend, 1962, Table 14). This suggests that, in line with demographic trends, the average age of care home residents has increased substantially over the last 50 years.

So how has the provision of care homes changed in response to these demographic changes?

Changes in residential care provision

Table 2.1 shows a number of significant changes in the provision of places in care homes and long stay hospital beds in the UK since 1970.[1] First, the total number of places more than doubled between 1970 and 1995, after which they started decreasing.[2] Despite this doubling, the provision of long-stay places per head of population aged 85 and over has decreased overall, although the provision of residential and nursing home places for this age group has increased slightly.[3] The overall decline is in part due to the shift towards 'community care' policies, including the development of various forms of supported housing. Nevertheless, as we demonstrate in this chapter, expenditure on care home provision has remained high and dominated social care budgets for older people. Second, it is clear from Table 2.1 that the late 1980s and early 1990s were a turning point in several respects: both long-term hospital provision and local authority residential care provision declined dramatically during this period, but this decline was more than compensated for by the increase in places in private sector residential and nursing home provision. Whereas the majority of places in residential and nursing homes were provided in 1970 by the public sector, in 2007 the situation was the reverse; three in four places were provided by the private sector and the voluntary sector provided more than the public sector. In addition, demonstrating just how far the boundary between NHS health care and means-tested social care has shifted towards the latter, the balance between nursing home places and residential care places has changed quite dramatically. Whereas in 1970 the ratio of nursing home places to residential care places was 2:17, in 2007 it was approaching 2:3. What is interesting, and something that commands less attention from social policy commentators is the stability of the voluntary sector in the provision of residential and nursing care with relatively modest increases in the number of places it provides and the share of the market it commands.

Within the broad context of the UK, we cannot ignore developments in the four nations particularly since devolution at the end of

Table 2.1 Long-stay places for older and physically disabled people (000s) by sector, UK, 1970 to 2007

	Residential Homes			Nursing Homes		NHS Hospitals		Total
	Local authority	Private	Voluntary	Private	Voluntary	Geriatric	Psycho-geriatric	
1970	108.7	23.7	40.1		20.3	52.0	23.0	267.8
1975	128.3	25.8	41.0		24.0	49.0		268.1
1980	134.5	37.0	42.6		26.9	46.1		287.1
1985	137.1	85.3	45.1		38.0	46.3		351.8
1990	125.6	155.6	40.0	112.6	10.5	47.2	27.0	518.5
1995	85.1	169.3	56.7	193.4	17.9	29.2	18.2	569.8
2000	59.7	185.0	54.5	186.8	18.0	14.4	11.7	530.1
2005	40.7	179.2	52.8	160.5	15.0	12.1	8.6	468.9
2007	37.5	183.7	50.4	161.2	15.9	11.1	7.1	466.9

Source: Adapted from Laing & Buisson (2008, Table 6.2).

the twentieth century.[4] Townsend's research was conducted in England and Wales and, although comparative data for the two are hard to acquire, there is some evidence of a stronger commitment to public sector provision in Wales. For example, within the overall decline in care home places since 1995 (see Table 2.1) and the corresponding home closures over that period there have been some interesting differences. Banks et al. (2006) found that, between 1991 and 2001, local authority homes decreased by over 50 per cent in England whereas in Wales they decreased by just 30 per cent. Furthermore, whereas the level of provision in the private and voluntary sectors increased slightly over this period in England, in Wales it decreased by ten per cent (ibid.), suggesting that the private care market is not as robust in Wales.

Before moving on to explain these changes, it is important to note that, at the same time as care homes were becoming a predominantly private sector business, council housing was being sold off to tenants or transferred in bulk to non-profit organizations.[5] Consequently, the proportion of owner occupiers increased substantially and likewise the proportion of people liable to have assets taken into account when being assessed for the provision of residential care. In 2005, just under three-fifths of people (58 per cent) over pensionable age owned their home outright (ONS, 2007, Table 10.8) compared to just over two-fifths (42 per cent) in 1980 (Mackintosh et al., 1990, Table 15). People over pensionable age were also more likely than younger people to be renting from the social sector, over a third of pensioners compared to under a quarter of non-pensioners in 2005 (ONS, 2007, Table 10.8), a substantial proportion of whom would be renting 'specialized' accommodation for older people such as bungalows and flats in sheltered housing schemes.

The general trends in care home provision over the last 50 years therefore have been the shift from public to private provision, the shifting boundary between free health care and means-tested social care and the shift of costs away from the state towards the individual 'consumer'. These developments in care are a very far cry from what Townsend recommended or possibly envisaged. To understand why this is so, we need to place them within the broader context of policies for older people since the 1950s and the founding of the welfare state.

The welfare state: The early years

The post–Second World War legislative framework within which services for older people were provided was established in England and Wales through the National Health Service Act 1946 and the National Assistance Act 1948. The latter officially abolished the Poor Law. This new framework established the distinction between the role of the NHS on the one hand and local authorities on the other in caring for older people. Indeed, since this time, boundary disputes between health and social care have been a dominant theme in both policy and practice. In addition, the Housing Act 1949 extended the powers of local authorities to provide housing, including accommodation for older people. During the first 20 years of the new welfare state, there remained a heavy reliance on residential care provision for older people. Domiciliary and day care services developed only gradually, their growth being hampered by shortages of staff, lack of finance and the low priority accorded to these kinds of services at national level (Sumner and Smith, 1969).

It was the intention of the post-war Labour government to change the face of residential care for older people. The Minister of Health, Aneurin Bevan, announced in 1947 the provision of special homes catering for 25–30 older people unable to look after themselves; the workhouse was to go, replaced by 'sunshine hotels' (Means and Smith, 1983). Between 1948 and 1959, l045 homes were opened, but only 160 were purpose-built (Townsend, 1962, p. 37). The vast majority were adapted premises, the limitations of which, as Townsend's research demonstrated, soon became apparent. It was not until the end of the 1950s and in the early 1960s that proposals were put forward to ease the provision of non-institutional services. At the same time, restrictions on capital projects were lifted and the provision of residential care began to accelerate, continuing to dominate local authority expenditure on services for older people.

The idea of providing support through care in the community rather than in institutional settings was given a boost by the recommendations of the Royal Commission on the Law Relating to Mental Illness and Mental Deficiency (1957) and the ensuing Mental Health Act 1959. However, while the government of the early 1960s committed itself to a major rundown of mental hospitals, this became linked to long-term planning for hospitals in general. The government sent a circular to local authorities pointing out that the plan for the development of the hospital service was complementary to the expected development of preventative and care services in the community. So, local authorities were asked to review their health and welfare services and to draw up ten-year plans. In respect of older people the circular suggested that

services for the elderly should be designed to help them to remain in their own homes as long as possible. Residential homes are required for those who, for some reason, short of a need for hospital care, cannot manage on their own.

(Ministry of Health, 1962)

The plans had, at best, a modest impact on the shift of priorities from residential to non-institutional care, partly because of the problem of so-called bed blocking (Hall and Bytheway, 1982). There was pressure from hospital authorities to increase the capacity of local authorities to provide residential care for frail, but not acutely ill, older people. Between 1959 and 1969, reflecting the relaxation of financial restrictions on the provision of residential accommodation, the number of local authority homes increased by 67 per cent and of private and voluntary homes by 45 per cent. While the latter provided more homes than the former, they catered for substantially fewer residents. However, local authorities were still accommodating older people in large institutions; in 1965 it was estimated that 27,000 people were still living in over 200 former workhouses (Ministry of Health, 1966, p. 28). So, despite the policy rhetoric and awareness of the growing costs of residential care, the period from 1964 to 1970 saw no great expansion of domiciliary services (Means and Smith, 1985, p. 219). At the same time, many local welfare authorities were failing to keep pace with the demand for residential care. Nearly a third of authorities in England and Wales (excluding the Greater London area) had less than the Ministry of Health's recommended lower limit of 15 local authority places per thousand elderly population. Although voluntary and private homes compensated to an extent for these low levels of provision, this was more marked in areas to which older people migrated following retirement (Sumner and Smith, 1969, p. 65).

Although, as mentioned in Chapter 1, there was a strong critique of institutional care emerging in the 1960s, including Townsend's, other sources presented a more benign view. An official survey, published in 1968, concluded that 'the gloomy picture of old people's Homes being inhabited by masses of unhappy, discontented residents is not supported' (Harris, 1968, p. 50) and the report of a committee on the staffing of residential homes anticipated an increased demand for residential care because the new homes being built were

delightfully planned and well equipped; if they further develop to give their residents help in living an interesting and sociable life it could easily happen that a much larger proportion than at present would choose to live in them rather than maintain their separate homes.

(Williams, 1967, p. 114)

Indeed, the late 1960s and early 1970s witnessed something of a building boom for new residential homes to replace the old workhouses and the unsuitable converted properties bought for use as homes for older people in the 1950s (Means et al., 2002, pp. 50, 52).

The reorganization of social services in the early 1970s, broadly following the recommendations of the Seebohm Committee (1968), held out the promise of change. However, the ten per cent growth targets soon had to be reduced when the recession of the mid-1970s began to develop (Walker, 1985). Both capital programmes and revenue expenditures had to be curtailed, affecting routine maintenance as well as upgrading or new building. Despite this, the number of residential homes continued to increase in the 1970s, partly as a result of demographic trends as well as the result of inadequate alternatives in the form of domiciliary and/or day care services. The resulting lack of policy shift led Bosanquet (1978) to comment that 'we are spending far more keeping residents in homes than we are on the elderly in the community in face of evidence that the great majority even of the severely handicapped housebound live in the community' (p. 120).

The restrictions on expenditure, linked to the growing concerns about future demographic trends, the increasing dependency of residents in homes and its implications for staffing and the quality of care (Means et al., 2002, pp. 54–8), were reflected in a discussion document published by the Labour government in 1978. *A Happier Old Age* (DHSS/Welsh Office, 1978) stated that gradual growth of residential and community-based services could no longer be guaranteed. Instead, there would have to be an extension of family and community support, plus the development of a wide range of housing options for older people, going beyond the provision of sheltered housing (Means et al., 2008, p. 165). A year after publication, a Conservative government was elected and this was to bring about some very major changes in services for older people.

The Conservative government's White Paper, *Growing Older* (DHSS, 1981) went even further in arguing that

> the primary sources of support and care for elderly people are informal and voluntary. These spring from the personal ties of kinship, friendship and neighbourhood. They are irreplaceable. ... Care in the community must increasingly mean by the community.
>
> (p. 3)

The image of residential care as a 'bad' option for older people reinforced the appeal of non-institutional care based on personal ties. The argument for placing less emphasis on residential care was boosted by research in the 1980s that confirmed Townsend's picture of isolation, depersonalization and lack of privacy. Booth (1985) found strong evidence for his hypothesis of 'induced dependency' in his longitudinal study of 7000 residents in 175 homes and Wilkin and Hughes (1987) concluded from a study of 60 residents in six local authority care homes that they moved into a state of total dependence. However, the argument was essentially finance led. Frail and sick older people were seen as a worrying 'burden' on the state and the notion of replacing residential homes with 'sophisticated packages of domiciliary services' (Means and Smith, 1985, p. 359) or the development of a complex of residential flatlets (Peace et al., 1982) was outweighed by concerns about pressures on the public purse. The real desire was to promote home-based care without spending too much money (Means and Smith, 1985, p. 361). Nevertheless the 1980s did see an increasing interest in the development of sheltered housing as an alternative to residential care with the focus being on provision by housing associations rather than local authorities (Butler et al., 1983). Furthermore, the Wagner Report (1988) noted that it was difficult to pin down the notion of residential care because its boundaries 'are constantly shifting with the emergence of new forms of provision' (p. 17).

Privatization

The overall concern for reining in public expenditure sits oddly with the then Conservative government policy in the early 1980s to encourage the development of residential care in the private sector. In 1983, the DHSS amended the supplementary benefit (income maintenance) regulations to make it easier for the low-income residents of private and voluntary homes to claim board and lodging allowances from the social security system.[6] This subsidy was based solely on financial entitlement and not at all on an assessment of need for this kind of care. Consequently the level of provision in the voluntary and private sectors (particularly the latter) trebled between the late 1970s and 1991. Likewise, over the same period, the social security bill for board and lodging allowances rose from £10 million to £1872 million, and by 1993, the year of implementation of major reforms of community care, the number of claimants was over 280,000 and the bill amounted to £2600 million. As both the Audit Commission (1986) and the Firth Committee (1987) had concluded, these financial arrangements provided a perverse incentive towards the use of residential care by local authorities, rather than the provision of home-based care. In addition, in order to access social security monies to support those in need of care, cash-strapped local authorities began transferring ownership of their homes to the independent sector. One of the aims of the NHS and Community Care Act 1990 was to remedy this situation. Following its implementation, the social security route into residential care was closed off and those requiring financial assistance had to turn to their local authority. Eligibility for financial support was to be determined not only by a means test but also by an assessment of need for care. Effectively entry into residential care for those with limited means was transferred from an open-ended central government department budget to ring fenced (and limited) local authority budgets. The new legislation also introduced a split between purchasing and provision so that local authorities would move away from direct service provision and develop instead an enabling and regulatory role. Indeed, there were specific disincentives for publicly owned provision in the new financial arrangements and one consequence was that it continued to be attractive for local authorities to contract out their residential care homes to the independent sector. As a result, as Table 2.1 shows, the balance of ownership between the public and private sectors continued to shift away from the former towards the latter.

The requirement to use the funding transferred from the Department of Social Security to support residents in the independent sector, together with restrictions on the level of fees paid by local authorities and an increased emphasis on standards, resulted in a struggle to survive on the part of proprietors of smaller homes. The good old days of 'fairly secure fiscal environments' were over (Knapp et al., 2001, p. 292). Stimulating the market in residential care was proving to be less than straightforward (Means et al., 2008, p. 71). The phenomenon of corporatization was becoming apparent. Between 1988 and 1997, the major providers (defined as organizations with three or more homes) more than doubled their share of the for-profit care home market. By 1999 they controlled over 30 per cent of the market. The process of small home closures, mergers and acquisitions was resulting in the concentration of ownership of long-term care provision into fewer hands (Holden, 2002; Darton, 2004).

Netten et al. (2005) examined the increased rate of home closures between 1998 and 2000 resulting in a net reduction in capacity, mainly as a result of the closure of smaller residential and nursing homes. A further study of home closures between 2002 and 2003 reinforced this finding (Dalley et al., 2004). In summary, the reasons for closures were financial: the perceived costs of adhering to the new standards, increasing staff costs and difficulties in recruitment, inadequate levels of fees paid by local authorities who were funding two-thirds of care home places, reduced demand for places leading to under-occupation and financial shortfall, increased costs associated with higher levels of dependency among the residents, changing values of premises in the property market such that a home was worth more empty than as a going concern, consequent problems in selling the home as a business. All these financial concerns led to a lowering of morale among providers and a loss of motivation (Netten et al., 2005). Laing & Buisson's (2007) report on the state of the market in the UK for the care of elderly people noted that 269 private and voluntary sector homes had closed in the year up to April 2007.

The growth of provision by the independent sector throughout the 1980s raised concerns about variation in the quality of care, the consequences of the profit motivation and the use of public funds without public accountability. Consequently, new legislation[7] regarding the regulation and inspection of private and voluntary homes, the Registered Homes Act 1984, was passed together with the publication of a (non-mandatory) code of practice for residential care, *Home Life* (Avebury, 1984). One of the consequences of these developments was a concern about the quality of care in local authority homes compared with the requirements being placed on the independent sector. The NHS and Community Care Act 1990 addressed this issue by requiring local authorities with social services responsibilities to establish 'arms length' registration and inspection units to assess homes in all sectors, using identical standards, which included individual care plans, a complaints system and a charter of rights for residents. The new arms length inspection units developed somewhat erratically, however, and they were not consistent in how they undertook the regulatory role (Day et al., 1996). This led to alternative models being put forward for the future of regulation, including the development of a national inspectorate (DH/Welsh Office, 1995).[8]

Modernization

The arrival of a Labour government in May 1997 heralded a 'modernisation' strategy (DH, 1998). Its focus on the notion of 'Best Value' shifted the emphasis from who should provide the care (the thrust of the policy initiative on residential care by the previous Conservative government) to placing it firmly on quality and improvement. One of the new government's manifesto commitments was the setting up of independent inspection and regulation for residential (and domiciliary) care. A specific set of standards for residential and nursing homes was drafted by the Centre for Policy on Ageing (DH, 1999), the same body that produced the 1984 *Home Life* document. While these were welcomed by statutory agencies and by organizations representing service users, because of the cost implications, they alarmed the providers of care homes, predominantly in the voluntary and private sectors. As a result, the final version of the standards (DH, 2001a) extended the deadline for achieving them until April 2007. However, this was not enough. Responding to further pressure, including significant numbers of home closures, the government made further amendments allowing care homes that opened before April 2002 not to have to meet some of the physical standards, such as room size (DH, 2002). In addition to the physical standards, other pressures on the providers of residential care included local authority pricing policies, the introduction of the national minimum wage in April 1999 and the European Union working time directive in October 1998. Despite these pressures, by 2005 it was reported that the proportion of homes in England meeting the standards had risen from 26 per cent to 48 per cent in the previous year (DH, 2005a, p. 61). However, the compromise on standards suggests that improving the quality of care is not wholly compatible with marketization.

Responsibility for monitoring quality was addressed in the Care Standards Act 2000 which, in replacing the Registered Homes Act 1984, removed the inspection/regulation role from local authorities and placed this task on a new body, the National Care Standards Commission (NCSC). Its remit naturally included key concerns such as choice of accommodation, size of rooms, access to health and social care, staffing levels, complaints systems and the availability of social activities. The NCSC started its work in April 2002 and before the end of that month was informed that in April 2004 it was to be merged into a new body, the CSCI in England, combining NCSC responsibilities with those of the joint reviews by the Social Services Inspectorate (SSI) and the Audit Commission. The announcement of this merger was accompanied by the comment that the possibility of a further merger between CSCI and the Healthcare Commission in England was under review (CSCI, 2008, p. 15). The new Care Quality Commission (CQC) started work in shadow form in 2008 and took over fully in April 2009. In Wales in 2007, the Care Standards Inspectorate for Wales (CSIW) was merged with the Social Services Inspectorate for Wales to become the Care and Social Services Inspectorate for Wales (CSSIW).

A further aspect of the modernization agenda related to the NHS (Secretary of State for Health, 2000a). Radical changes to health policy in both England and Wales to enhance primary care rather than being hospital driven had been outlined in two White Papers (DH, 1997; Secretary of State for Wales, 1998). Each health authority in England was required to produce a health improvement programme (HIMP) reflecting both local and national priorities. These priority areas had to have their own national service framework (NSF), and one of these frameworks was for older people (DH, 2001b). An NSF for older people in Wales followed later (WAG, 2006). The focus of the English NSF, however, was mainly on acute health or community-based services. Residential care was only mentioned in the context of 'intermediate care': 'Our older people will have access to a new range of intermediate care services at home, or in designated care settings, to promote their independence and to prevent premature or unnecessary admission to long-term residential care' (DH, 2001b, p. 13). The focus, therefore, was not centrally on the quality of life for people in residential care. Rather it was concerned with early (presumably not premature) discharge from hospital. This led Means et al. (2002) to comment that

> [i]t is hard to avoid the conclusion that what is being demanded is the speedy removal of older people from hospital. One hopes this will occur through a greater investment in genuine rehabilitation services, rather than just under the guise of rehabilitation. It would be too easy for much private residential and nursing home accommodation to be redefined as short-term rehabilitation beds, even though they might often lead on only to long term care in the same institution.
>
> (p. 165)

Clearly, the Labour government's modernization strategy gave residential care an essentially residual role, fuelled by a decline in provision. The House of Commons Health Committee (2002) concluded that, although nearly 30 per cent of delayed discharges from hospital were due to people awaiting transfer to a care home, the solution to such delays lay not in an increase in care capacity, providing 'more of the same' (p. 46), but in the development of alternative service models that more closely reflected the wishes of older people to remain in their own homes. 'Closure of care homes can act as a further spur to the development of care at home and other responses tailored to the assessment and preferred choice of individuals' (p. 40). Meanwhile, expenditure on residential care still exceeded that on day and domiciliary services for older people.

Paying for care

The issue of how long-term care should be paid for has been prominent in debates about policies for older people since the mid-1990s. In December 1997, the new Labour government established its promised Royal Commission to come up with recommendations for a sustainable system for funding long-term care. Its report was published 15 months later (Sutherland, 1999a). Apart from the general drive to develop policies to allow people to remain in their own homes, the decision to establish this inquiry was fuelled by concern about the depletion of older people's capital resources, particularly home equity, as a result of expenditure on residential or nursing homes. The majority report of the commission concluded that there should be no distinction between funding for long-term health care and personal care,[9] and that the care should be paid for out of public funds. The minority report argued that this proposal would require a huge transfer from the private to the public purse which would not increase spending on services for elderly people 'by a single penny' (1999a, p. 113). The latter view was accepted by the Labour government (Secretary of State for Health, 2000b) which made nursing care free in care homes in England, but social (personal) care was to remain means tested and chargeable. Hence disputes between what constituted nursing care and what constituted social care continued (Blakemore and Griggs, 2007). Furthermore, payments for nursing care were made at three levels following an assessment of need, leading to further inequities. While Wales also followed the minority report, it introduced a flat rate payment for those requiring nursing care.

In September 2003, the signatories to the majority report of the Royal Commission on Long Term Care published a review of developments since March 1999, the date of their original report (Sutherland et al., 2003). They concluded that the case for free personal as well as nursing care remained incontrovertible, arguing that 'intimate care that directly involves touching a person's body ... was sufficiently connected to "health care" to justify state support' (p. 4). They challenged the appropriateness of the definition of nursing care pointing out that this excluded the care provided by health care assistants 'who in practice deliver most of the care of older people in care homes' (p. 6). Intermediate care was essentially short term and did not address the central issue of long-term care. Free nursing care had not eradicated the problem of bed blocking, and social care remained underfunded.

Social care is neither a luxury nor an added extra. It is vital to the well being of very many older people. Underfunding impacts directly on their lives. It leads to unacceptable rationing of social care.

(Sutherland et al., 2003, p. 8)

The Commissioners therefore asked the government to reconsider its policy on care funding and declared that they would 'continue to press the case ... on the grounds that it is a just, principled and affordable way of meeting a pressing social need' (Sutherland et al., 2003, p. 16).

In 2005, the Office of Fair Trading published a market study of care homes for older people, triggered by a complaint made by *Which?* in 2003 on behalf of the Social Policy Ageing Information Network. It focused on the consumer's perspective including that of the 30 per cent of care home residents who were self-funders. It found that 54 per cent of the homes were single home businesses (2005, p. 41, quoted in Scourfield, 2007, p. 163) and the core issue was whether older people were provided with reliable information about a home and its providers, their rights within the home and a clear, fair contract to help them come to a good, initial decision at a time when they were likely to feel particularly vulnerable. The report was critical on all counts, while acknowledging individual examples of good practice. It made wide-ranging recommendations for the creation by government of a one-stop shop to obtain information about care for older people, the development of pilot advocacy services and for improvements in practice by local authorities as information providers, not least on funding matters. It also recommended that access to reports by inspection bodies, in their capacity as monitors of information clarity, accuracy and complaints systems, should be improved; that care home providers should issue inspectors' reports to potential residents and/or their families as well as to existing residents; that prospective residents should be given current prices in writing by a home prior to any decision to take up residence and that all residents should be issued with a written contract. Interestingly, the report acknowledged that self-funders were often paying a higher price for the same service that local authority funded residents were receiving, but it argued that as local authorities were 'bulk buyers' this was acceptable.

Like the Royal Commissioners, the Wanless Report (2006),[10] the prime purpose of which was to influence the debate on the government's spending plans on services for older people, also raised concerns about the rationing of social care. It noted that tighter rationing criteria left people with less than severe needs with no public support and therefore vulnerable to admission to hospital or residential care as a result of a crisis. It recommended a 'partnership model' of funding whereby everybody in need would be entitled to an agreed level of free care, after which individual contributions would be matched by funds provided by the state up to a further agreed limit. Those on benefit could use these funds to pay for their contributions. Above this higher limit, people could use their own resources to top up the service(s) needed. It was calculated that this model would result in a doubling of the percentage of gross domestic product devoted to social care from 1.1 per cent in 2002 to 2.0 per cent by 2026. Also in 2006, the Joseph Rowntree Foundation published the results of its programme on paying for long-term care (Croucher and Rhodes, 2006). It characterized the existing system as inadequate, unfair and incoherent. It suggested a number of measures that could be undertaken in advance of a radical overhaul of the current system, such as better arrangements for releasing equity in homes and higher capital thresholds so that more people could be eligible for local authority support. Both the Wanless Report and the work published by the Joseph Rowntree Foundation were reflected in the Green Paper, *Shaping the Future of Care Together*, published in July 2009 (DH, 2009).[11]

Personalization and residential care

The Royal Commissioners had to wait 18 months for a response to their review in the form of a consultation document (DH, 2005a), the central message of which was to give individuals, their friends and families greater control over the way in which social care supports their needs. This included much wider usage of 'direct payments' arrangements, but in the context of developments over the next 10–15 years being cost neutral for local authorities. In respect of residential care, this vision of control by service users introduced the idea of giving individuals the right to request *not* to live in a residential setting (hardly a ringing endorsement of the service provided by care homes). Illustrating yet again the tension between provider and user preferences, this right 'would require service providers to make explicit the reasons behind their decision to recommend residential care, including cost considerations' (2005a, p. 32). The subsequent White Paper for England on the future of health and social care (DH, 2006) emphasized the need to improve services to people in their own homes. As far as the future of social care was concerned, self-directed care and support was again emphasized through the enhanced take-up of direct payments and the piloting of individual budgets. The latter would be underpinned by combining several income streams, none of which could be used to pay for residential care.

In 2005–6 gross expenditure on residential care was nearly £5000 million, whereas the gross costs of direct payments were £274 million (CSCI, 2008, p. 20). If the idea of personalized care was to include better support for people to stay in their own homes, there was a long way to go. The number of permanent admissions to residential care in England of people aged 65 dropped between 2005–6 and 2006–7 by 4900 (CSCI, 2008). Despite this, as noted by CSCI, the shift of expenditure from residential care to community services was relatively small: between 2001 and 2006, there was an increase of only one per cent in the proportion of expenditure on services committed to help older people living at home. This was explained as being the result of the increasing numbers of very old people with local authorities having to look after more people with increasingly complex needs. Therefore it was difficult to shift resources to community services, which, in any case, would be costly for people with high levels of need, so services were being reduced for people with less complex needs. In 2005–6 just over a half of local authorities in England restricted the provision of services to those deemed to be at a substantial or critical level of risk and this was expected to rise the following year (CSCI, 2008). With the increasing numbers of older people, the care home population is expected to rise after 2012 having fallen year-on-year since 1993 (*Community Care*, 2007).

This then was the situation at the end of 2006 when we completed our fieldwork. The story is incomplete however without giving some attention to the parallel developments in research following the publication of *The Last Refuge*.

Changing research agendas

As we discussed in Chapter 1, *The Last Refuge* was one of a series of critical studies published in the 1960s that have been described as the 'anti-institutional literature'. The impact of Townsend's findings, methods and recommendations on subsequent research into residential care, has been quite substantial. Although he recommended that residential care should be phased out over a period of five to ten years, he also proposed 'communal accommodation', providing short-term care or offering a choice to 'physically or mentally incapacitated persons' (1962, p. 420). He issued some very detailed recommendations on the environmental and material conditions of such homes (1962, pp. 420–3). His concern was that, as far as possible, they should be modelled on 'the customs and practices of home and community life'.

For example, he recommended siting homes near to shops and local amenities; that the majority of rooms should be single; that small sitting rooms and dining rooms should be available in each home; and that residents should be encouraged to participate in ordinary daily domestic activities, and to manage their own pension books. His recommendations were aimed at maximizing 'individuality, freedom, self-determination, the right to occupation and the opportunity to form new social relationships, as well as preserve old ones' (p. 419).

Following publication of *The Last Refuge*, a burgeoning research literature on residential care for older people has emerged (see, for example, the reviews of Wilkin and Hughes, 1980; Davies and Knapp, 1981; Judge and Sinclair 1986; Sinclair, 1988; Sinclair et al., 1990; Elkan and Kelly, 1991). Indeed, by the early 1980s some social gerontologists had become critical of the attention and resource being devoted to what, in the wider context of ageing research, they considered a minority issue. Our objective here is to identify some of the key developments in the history of research in this area since Townsend undertook his highly influential study nearly 50 years ago, and to better understand the context of our own research and the strategies and methods we have adopted.

In 1962, the year that saw the publication of *The Last Refuge*, a Local Authority Building Note issued by the government advocated the building of purpose-built homes for between 30 and 60 residents in locations that facilitated community integration and providing residents with an 'informal and homely atmosphere'. Over the following ten years, informed by case studies undertaken by architects (for example, Ministry of Housing and Local Government, 1966), there was extensive debate as new homes were designed and built. This generated growing concern over some of the contradictions entailed in providing a homely atmosphere for up to 60 people. Concurrently, research located in the NHS reflected a growing interest in the relationship between hospitals and residential homes in providing effective care for older people in need of long-term medical treatment (for example, Kay et al., 1964).

A key concern of care home providers, prompted by the anti-institutional literature, and in particular by Goffman (1961), was the 'block treatment' of residents. Hanson, for example, a director of social services, took this up in order to develop ways of creating environments and practices which would enhance the quality of life in residential homes. During May and June 1969, he spent time in a home observing its organization and routines and, subsequently, he published an influential report (Hanson, 1972). Reflecting the same concern to mitigate the institutional features of large homes, a revised Building Note was issued by the government in 1973 to advise local authorities to divide the residents of homes into smaller family-like groups (DHSS/Welsh Office, 1973). This, it was argued, would allow them more privacy and self-determination.

The 1970s saw an increasing involvement of social researchers. At this time, most studies were conducted in homes in the public sector. This was because it was readily accessible to researchers. Townsend himself acknowledged that local authorities were 'particularly generous in affording every possible facility' to him (1962, p. 9). Lipman (an architect) and Slater (a psychologist) undertook substantial observational research in homes where areas of staff activity were separated from those of residents. They demonstrated how this reduced opportunities for staff surveillance and intervention, and how such separation promoted the residents' needs for independence, activity, privacy and integration (Lipman and Slater, 1976, 1977). They also focused on the contribution of staff attitudes and behaviour to the institutional environment, and on the nature of interactions between staff and residents (Lipman et al., 1979). Jenks (1978) drew attention to the distances some residents had to cover within the new purpose-built homes, arguing that these restricted activity and reduced mobility and independence. The Wyvern Partnership (1977) revealed how confusion could be built into those homes that used standard unit designs that were repeated without variation in the spatial layout.

Alongside this work on design and layout, there was growing interest in the residents themselves and their needs (Carstairs and Morrison, 1971). Whereas Townsend had sought to measure the capacity of residents for self-care, research in the 1980s turned to the development of broader indicators of function and dependency (Wilkin and Thompson, 1989). Booth's study of 175 homes examined the relationship between regime and induced dependency (Booth, 1985). Research also focused on the reasons for admission (Neill et al., 1988) and on ways of managing the mix of 'lucid' and 'confused' residents (Evans et al., 1981). Adopting a more consumerist approach, Willcocks and colleagues devised ways of eliciting the views of residents (Willcocks et al., 1987). All of these researchers benefited from the interest of policymakers in government departments and the access offered by local authority managers.

With the shift away from public sector provision in the 1980s and the development of a mixed economy of care, the focus of research began to change. Large-scale studies involving detailed observational work were no longer possible and the expanding private sector, particularly in England, was more difficult to access. Research became more regional, such as the study of the growth of private sector care in Norfolk (Weaver et al., 1985) and Devon (Phillips et al., 1988) and of the quality of residential care in Scotland (Bland et al., 1992). The focus of attention, therefore, turned not only towards developments in private sector provision but also to matters of regulation and inspection and ways of assessing and monitoring care provision and the quality of life. A substantial body of research was produced, such as that emanating from the Centre for Environmental and Social Studies on Ageing (see, for example, Kellaher et al., 1988) and the Personal Social Services Research Unit (see, for example, Netten, 1993; Netten et al., 2001; Bebbington et al., 2001). During the 1990s, with increasing pressures on the private sector and the shift of costs towards the 'consumer', research followed that developed 'inside quality assurance' (Peace and Kellaher, 1993) and assessed the impact of charging polices on older people entering residential care (Wright, 2003; see also Sutherland, 1999b). More recently, following an increase in home closures in the late 1990s, research has turned to the consequences of this for residents and those responsible for their welfare (Netten et al., 2005).

Attention was also moving towards the consequences for residential care of an increasingly frail population of residents. During the 1990s, dementia in particular became a growing issue and investment in research in this area expanded substantially. Through the 1970s and 1980s, dementia had largely been the territory of bio-medical research, but Kitwood (1997), among others, challenged this approach. Through the campaigning of the Alzheimer's Society, dementia became a political issue. The consequence was that a dementia perspective was cast upon many issues of relevance to residential care, such as design and assistive technologies (Judd et al., 1998; Marshall, 2001), communication and reminiscence (Gibson, 1994; Schweitzer, 1998; Killick and Allan, 2001) and, more generally, the development of person-centred care (Kitwood and Bredin, 1992; Kitwood, 1997; Brooker and Surr, 2005) and the rights and experience of people with chronic illness (Downs, 1997; Clare et al., 2008).

In addition, following the lead of research into child and domestic abuse, research into elder abuse developed during the 1990s (Bennett, 1990), and this included a focus on the risk and protection of residents of care homes (Garner and Evans, 2000). This became linked to the development of broader policies regarding the protection of vulnerable adults (POVA) and the publication of guidelines for provision and practice (DH, 2000).

Throughout the 1990s there was concern over the roles and resourcing of residential care homes and nursing homes, reflecting the changing balance between social care and nursing care. With the expansion of nursing research in universities, residential care has become very much a part of the nursing research agenda for the twenty-first century. Some research has focused on relocation issues and moves from hospital to care homes, as well as moving between homes (for example, Reed et al., 1998; Morgan et al., 1997). Other research has focused on the utilization of primary care services in care homes (Goodman and Woolley, 2004), medication (Corbett, 1997) and ways of improving the quality of care more generally (Froggatt et al., 2008). The Senses Framework, at the heart of relationship-centred care (Nolan et al., 2001), has been central to the development of practice through the 'My Home Life' Initiative (www.myhomelife.org.uk). Importantly, research has also begun to address issues of finitude and approaches to end of life care in care homes (Katz and Peace, 2003; Froggatt and Payne, 2006). Meantime, the focus of researchers more interested in housing has shifted towards policy developments such as the provision of 'extra care' housing.

In the early 2000s, there has been a greater emphasis on actively involving older people in research (Peace, 1999). This is seen as a way of empowering service users and, connected to this, of ensuring that funded research is both useful to 'users' of research and ethical (DH, 2005b). The ESRC, for example, in funding projects such as ours, requires researchers to actively engage with older people. Care home residents are no longer regarded simply as research subjects but rather as participants. In addition, new approaches to qualitative research and the analysis of qualitative data have emerged, moving away from the positivist paradigm to a more constructivist and reflexive one, where the researchers and researched are seen as co-producers.

It is within the context of these developments in research that we made plans to revisit Townsend's homes and to replicate as far as possible his method of inquiry.

3
The Study Design and Methods

Designing a revisiting study is not without its challenges and in this chapter we detail how we approached them.

As mentioned in Chapter 1, having conducted a survey of residential care homes in all local authorities in England and Wales, Townsend then drew a stratified random sample[1] of 180 homes, 173 of which he visited. As Table 3.1 shows, they included: former PAIs (ex-PAIs)[2] which were still being used to accommodate large numbers of older people; other local authority homes, five of which were purpose-built and the remainder converted properties; voluntary and private homes. Townsend organized his data collection and analysis around these four types of homes and *The Last Refuge* includes a chapter on each.

Our first task was a visit to the archive in order to draw up a list of the homes Townsend visited. *The Last Refuge* data are filed in Boxes 36 and 37 of the Peter Townsend Collection. We found Townsend's list, neatly handwritten, in one of the files in Box 37 which contained an assortment of questionnaires. It included the names of the homes and their locations (but not specific addresses), for example, 'London, Southern Grove'. The list was organized exactly as Townsend had sampled the homes (see Table 3.1) so that homes 1–39 were the ex-PAIs, 40–92 the other local authority homes, 93–131 the voluntary homes and 132–73 the private homes. The reports that Townsend and his colleagues wrote on each of these homes, which among other things provided more detailed clues regarding location, had not been archived according to provenance. Rather, they were organized alphabetically into 11 files (in Box 36) most often according to place but sometimes according to the name of the home. Each entry on the contents list of the files also included the 1958–9 tenure but the ex-PAIs and 'other local authority' homes distinguished by Townsend had been conflated into one category: LA (local

Table 3.1   Townsend's sample by tenure, 1958–9

Former Public Assistance Institutions	39
Other local authority homes	53
Voluntary homes	39
Private homes	42
Total	173

Source: Townsend (1962, Table 1, p. 10).

authority). We had to collate the two lists until we had a complete list, as per Townsend's, with the location of the report on each home in the archive noted against each home. By way of example, the contents list for file 5 included 'Lancashire, Highlands' and its tenure 'LA'. Checking this against Townsend's list, we found that number 14 on his list was 'Lancs., The Highlands, Wesham (JU)'[3] – one of the 39 ex-PAIs he had visited. We were then able to record on our own list that the home report on H14 was archived in file 5 of Box 36.

As a result of this exercise, we found that 40 of the 173 home reports were missing and nearly all of these were reports on the ex-PAIs or other local authority homes. Despite this, we were able to assemble a complete listing of the homes and therefore to design our revisiting study.

Central to our design were two related but discrete fieldwork studies:

- A tracing study through which we would find out what happened to the 173 homes visited in 1958–9.
- A follow-up study of those homes that had survived so that we could compare the homes then (1958–9) and now (2005–6).

It was essential for the latter therefore that through the tracing study we were able to distinguish between surviving and non-surviving homes. But this was not always straightforward.

Survivors or non-survivors?

In 2004, when we were preparing our research bid to the ESRC, we made a start on finding out how many of Townsend's homes had survived. The simplest way to do this was to put the name and location of the home into the website of the then NCSC (for England) or to contact the then CSIW or the organization that had owned the home in 1958–9. Initially this seemed straightforward, but when we did this for a home called Fairview,[4] we found that although the voluntary organization that had owned it in 1959 still owned a home of the same name, it was located some considerable distance away from the one described by Townsend in his archived report. We were unable to ascertain from 'head office' if there was any connection between the two homes. So, in order to investigate further, one of us set out by car to try to locate the original home.

The archived report indicated that the building had been acquired by the voluntary organization in the late 1940s to provide residential care for men returning from the Second World War who had nowhere to go or who had been wounded and needed care and attention. The report also contained the name of the village near which it was located. The old house was marked on the ordnance survey map so it was not difficult to find. As described by Townsend, it was still an imposing gothic Victorian hall situated in 62 acres of parkland with a drive nearly a mile long. The nearest village and shops were a good mile-and-a-half away. The house, it transpired, was now used as a retreat and conference centre and had not been a residential care home for nearly 20 years. However, we subsequently found out that, in October 1985, all the residents and some of the staff had been moved in two minibuses to become the first residents of a new, purpose-built home in the heart of a village 40 miles away, which had been given the name of the old home (Johnson et al., 2008).

This anecdote illustrates the simple point that care homes comprise a combination of people (staff and residents) and bricks and mortar (the building). Over time, in order to adapt to prevailing social and economic conditions, homes change or gradually transform into something different. Uncovering and explaining this process of change was, of course, the essence of our research. In deciding whether Fairview should be classified as a survivor, and therefore included in the followup study, the fundamental question we had to address was 'is this the home that Townsend visited in 1958–9?' We decided that it was. It was no longer in the same building, or even in the same place, but it had retained its institutional identity. Its building and its isolated situation had become unsuitable for the care needs of older people in the mid-1980s and beyond and so, as a community, it had moved elsewhere under the same ownership, taking its name with it. This was how the home had adapted to the changing context of care provision for older people.

In many cases, distinguishing between surviving and non-surviving homes was quite straightforward. St Peter's, for example, was, in 1958, a voluntary home owned and run by a Catholic Order. In 2006, it was still a voluntary home owned and run by the same Order and registered with CSCI to provide long-term care for older people. It has kept its name and is in the same building albeit with some adaptation and modernization over the years. This was an unambiguous survivor. In contrast, St Joseph's,[5] another voluntary home owned and run by the same Catholic Order was closed in 1978. The closure was planned and the residents were transferred to other homes run by the Order. The site was sold and redeveloped and the building was demolished in 1984. This was an unambiguous non-survivor.

Between these two extremes were homes whose status in terms of survival, like Fairview above, was more ambiguous. For example, some of the ex-PAIs which had been demolished had been replaced by 1960s purpose-built homes, built on the same site and sometimes assuming the same name as the old home. The residents had been transferred from the old to the new. We also considered these to be survivors. In other cases, however, following demolition, residents had been dispersed among several homes in the surrounding area. In these cases, there was no identifiable surviving home and no *continuing institutional identity*.

Two homes which we considered to be right on the cusp of the survival/non-survival continuum were Mardale and St Mark's. Both were local authority homes in 1958–9. Mardale closed and was demolished and the residents were dispersed elsewhere (probably in the late 1960s or early 1970s). The site remained a wasteland for some years. (It was extremely hard to find anyone who had any memory of the home). The site was eventually sold to the international healthcare company, BUPA, on which it built a new care home, which opened in 1983. St Mark's was a joint user institution[6] in 1958–9. The buildings have survived and still function as a hospital, but the last ward that accommodated long-term 'geriatric' patients closed in the late 1990s. In 2005, BUPA opened a new care home for older people on the hospital site taking the same name as the hospital. We categorized these homes as non-survivors on the basis that, apart from the location, there was nothing to suggest that these homes had an institutional identity that linked them to Townsend's research.

When the tracing study was complete, we had categorized 39 homes as survivors. The key indicators that we identified in determining whether a home had survived and sustained a continuing institutional identity are listed in Table 3.2. Nearly all the surviving homes were still in the same *location* – that is, on the same site. Just two purpose-built replacements were on a different site but both had retained their names and taken in the residents and staff from the former home. Most had had a

Table 3.2   Indicators of survival

Continuing features	No. of homes
Location	37
Population	33
Building	31
Name	26
Tenure	24
Provider	15

continuous *population* of residents that through deaths and new admissions since 1958–9 had gradually been transformed into the population of 2005–6. The population of six of the homes had at some time had a break due to rebuild, refurbishment or, in one case,because the home had for some time functioned as a children's home before returning once again to being an old people's home. Around three-quarters were still in the same *building* although all, as we explain in detail in later chapters, had been modified and/or extended in some way. Two-thirds had retained their 1958–9 *name* while the remainder had taken on different names when they changed owners. Slightly more had changed *tenure*, most notably, as one might expect, the former local authority homes. But nearly two-thirds had had changes of *provider* (that is the named owner). These were not only the local authority homes but also all of the private homes. It was the voluntary homes that were the most likely to be characterized by all these features and the former local authority homes that were the least likely (except for the two in Wales).

The tracing study

Although, through the data in the archive at the University of Essex, we had managed to assemble a complete list of the 173 homes, for many we did not have an exact address. 'The Rest Home, Sutton Coldfield' is a typical example. In this instance we were able to add from information available in the archived home report 'halfway down Chester Road'. Often such detailed additional information was not available. In particular, to trace the location and history of some of the small private homes, which may have closed or even been demolished in the 1960s, was likely to require a considerable amount of detective work.

Through the tracing study we aimed to provide fresh insights into the changing nature of accommodation and care for older people and to illuminate issues relating to the sustainability of residential care provision. By comparing the history of the closed homes with the history of those that have survived, we hoped to discover what factors enabled some homes rather than others to adapt to changes in policy and practice and to continue to provide a service from one decade to another. When did the home close? Why? What happened to the residents? What changes of ownership and use had there been? Was the building demolished and, if so, what now occupies the site? What kind of location was the home in and has that location changed significantly during the intervening years? It quickly became apparent from our own initial investigations that the kind of information required to answer such questions would be best obtained from someone with local knowledge: someone able to locate and visit the site, to consult local people and to examine documents in the local records office, such as local directories, newspapers, council minutes, planning applications, land registry documents and electoral registers.

We decided, therefore, that recruiting locally based volunteer researchers was not just the only viable way to proceed in practical and financial terms but also the most desirable in terms of outcome. Put simply, we needed to draw on local expertise, knowledge and experience. And older people in particular might have memories of homes now closed.

We put out our first call for volunteers in January 2004 when we were writing our research proposal for the ESRC. It was important at this stage to gauge the kind of response we might get. Our first port of call was the University of the Third Age (U3A). We chose U3A not simply because it is an age-based national organization but also because it is an educational organization, and we recognized the opportunity we could provide for its members to draw upon and offer us their expertise, as well as to develop their lifelong learning. Our call went into the U3A national newsletter and onto its website and we had a good response from members across England and Wales, largely but not exclusively through email. This reassured us that it was not a completely foolish idea that local people would volunteer their knowledge and expertise for what they saw as a 'worthwhile' project. At the same time, we also made contact with 20 local history associations across Britain. Interestingly their initial response was not as enthusiastic, perhaps because of the lack of a co-ordinated national structure. One local historian, however, who had also been a member of a Registered Homes Tribunal,[7] offered to do a pilot for us on two of the tracing study homes. He undertook this with the utmost rigour, producing detailed reports, one of which we were then able to use as a model in the *Information Pack for Volunteers* which we subsequently produced.

Table 3.3   Organizations through which volunteers were recruited

Organization	Volunteers
University of the Third Age (U3A)	42
Social Work History Network (SWHN)	14
Local history societies	9
Older People Researching Social Issues (OPRSI)	4
Older people's forums	4
Other	27
Total	100

Once we had obtained our funding from the ESRC, we extended our appeal to a broader range of organizations, as listed in Table 3.3. In addition to continuing to recruit through U3A, we recruited retired members of the British Association of Social Workers (BASW) who were members of the Social Work History Network (SWHN), set up in 2000. The network has regular meetings through which it explores and discusses matters relating to the origins of the social work profession, how it has approached particular problems in the past and what can be learnt from this. We also recruited four volunteers through Older People Researching Social Issues (OPRSI), a group of older people trained at the University of Lancaster as researchers, who subsequently set up their own research consultancy (Clough et al., 2004, 2006). A further four were recruited through older people's forums which had participated in another research project based at The Open University (Bytheway et al., 2007).

On the whole, organizations, particularly U3A, were more responsive in the south of England than the north. For those geographical areas that were less easy to cover we relied more heavily on individual contacts made through friends, local studies librarians and the recruitment page on our website. In addition, in order to recruit individuals for specific homes, some press releases were prepared for local newspapers. It was through these individual approaches that we attracted 27 additional local historians. Overall, 100 people volunteered their assistance, 79 of whom were active in tracing the history of Townsend's homes. Of the remaining 21 volunteers, five withdrew (one due to bereavement) and the rest either acted as contact persons, informants or reserves because there was no home in their area for them to trace.

Of the 79 who were active in tracing the history of homes for us, 62 completed a 'personal profile form'. Just under three-quarters (71 per cent) were women, just over three-quarters (77.4 per cent) were aged between 60 and 79 and a further ten per cent were aged 80 or more.

On the basis of information provided about previous employment, almost all the volunteers might be described as middle class: 44 per cent had worked in health, social work or social care occupations before retiring. A further 18 per cent had worked in education and eight per cent had been librarians or archivists. Only one had no formal educational qualifications. Sixty per cent had a first degree and/or a higher degree. Twenty-six per cent had no professional or vocational qualifications but of the qualifications held, 31 per cent were in social work or related occupations; 21 per cent in medicine, nursing and other health-related occupations; 14 per cent in teaching; and ten per cent in library and information services. The remainder (24 per cent) covered a range of qualifications including banking, management, engineering and surveying. In summary, the volunteers were on the whole highly educated and skilled and committed in some way to lifelong learning.

The personal profile form included an open-ended question about reasons for volunteering to take part in our project. Many of the responses were in keeping with the findings about volunteering in later life, generally, the opportunities and rewards it provides (Baines et al., 2006). What was distinctive about our volunteers, however, was the lifelong learning dimension. Our analysis of their responses showed three main reasons for volunteering. These were that the project provided the opportunity to contribute skills, knowledge and expertise; to learn new skills and acquire new knowledge; and that it was a timely opportunity for personal reasons.

Given the overall profile of our volunteers, it is hardly surprising that a frequently cited reason for volunteering was having skills, knowledge and experience that they thought they could bring to the project. There were many references to previous experience of social services, such as 'it sounded interesting and related to my previous work training residential care staff' or 'as a retired geriatrician, I have a lot of experience of elderly people's homes'. Eight referred specifically to Peter Townsend or *The Last Refuge*. For example, one said, 'Peter Townsend was something of a "guru" in my social work training', and another 'I recall reading *The Last Refuge* when I trained in 1966.' Apart from references to the experience of social services, there was also reference to skills in historical research: one mentioned voluntary work in the local records office and another to past historical research 'using university and city library archives'. Some indicated a strong personal awareness of age and the impact for them of retirement. For example, one man said that participation promised 'enjoyment of using old professional skills which, fortunately, have not rusted away!' and another 'I wanted to delay the atrophy of old professional and painstakingly acquired skills for just a little longer.'

In addition to professional skills and expertise, there were also some responses which indicated a specific knowledge of the locality or of the particular homes needing to be traced. Some mentioned having had a relative who had resided in the home they had been tracing. A former social worker had been involved in the assessment and placement of the residents when the home closed. Another, secretary of the local civic trust, had written an article many years back about the house in question. He also remembered it when it was a home. As he commented during a conversation on the telephone, 'it overlooked the Town Field and I remember seeing the old men sitting outside, but then someone removed the forms [benches] – typical of the council'.

Wanting to learn was a second important reason put forward for getting involved in our research. Many expressed this simply as 'it sounded interesting and worthwhile' but several specifically wanted to gain more research experience. There were a few who decided they wanted to read the original book and who devoted considerable effort to getting hold of a copy. One, for example, who lived in Dorset, managed through his local librarian to obtain a copy from a library in Oxfordshire. Apart from acquiring research skills, some indicated a growing interest in matters relating to ageing and later life from both a personal and professional perspective. One noted that the care of older people had become 'more important as the years go by' and another referred to the fact that her husband was now in a nursing home. A third said 'I'm getting older myself and this seemed so interesting.'

Thirdly, there were more personal reasons for getting involved, including responses from those who saw participation as a challenge, sometimes because they had time available which they wanted to use constructively. For example, one said 'I thought it would offer an interesting challenge' and another that they 'needed to regain an outside activity following a recent illness'. In addition, there were those who expressed loyalties either to The Open University, the University of the Third Age or their local library: people who were willing to volunteer on behalf of or be volunteered by an institution or individual they held in high regard.

Given the broad geographical spread and the fact that the volunteers were not all recruited at the same time but over an extended period, it was not feasible to set up a training course. Rather, each was individually briefed and supported. The briefing pack included information about the project and details relating to location (from the archive) of specific homes they had offered to investigate, guidance on how to search out relevant information, a letter of introduction and a standard form for completing their report together with two examples of completed reports. The pack also included an expenses claim form and a personal profile form. Each volunteer was allocated to one member of the research team who became their contact person and supporter. The volunteers were instructed not to knock on any doors without consulting us and to inform us immediately if the home turned out to be currently functioning as a care home. While we, the research team, had ethical approval and Criminal Record Bureau (CRB) clearance to conduct research in care homes, our volunteers did not and we were particularly concerned that they should not extend their inquiries beyond their remit. We discuss the involvement of older people as researchers further in Chapter 9 (see also Johnson et al., 2006).

As will become apparent in the next chapter, participation in our project offered the volunteers the opportunity to engage in the kind of detective work that characterizes research at its most exciting. They went to extraordinary lengths to follow trails and leads and to tie up loose ends. Most volunteers were investigating only one or a few homes so the history of each was given very comprehensive attention (see Appendix 2). This meant that not only did we receive very full reports but also that many volunteers were able to make a significant contribution to local historical knowledge.

When the tracing study was completed, we drew a random, stratified sample of the non-surviving homes, matched for size and tenure (as in 1959) with the surviving homes. This helped us to draw some conclusions relating to how and why different kinds of homes have adapted to changes in policy and practice (see Chapter 4).

The follow-up study

Prior to funding being obtained, we had already identified some 14 surviving homes which had kept their original names. So, although the follow-up study of the surviving homes was conducted alongside the tracing study, we were able to make a start by piloting our method by visiting three of the homes. As more surviving homes were discovered through the tracing study and added to the list, we had to devise a sampling strategy because we were funded to make follow-up visits to only 20 homes. We compiled our sample on the basis of maximizing diversity in terms of geographical location, tenure then and now, size and type of ownership. Through this strategy we aimed to capture the diverse ways that surviving homes have adapted to the changing social and economic conditions over the years.

Two of the 39 surviving homes were not included in the sampling frame because although both were still registered as care homes, they were no longer providing permanent accommodation and care for older people. One was providing respite or short-term care only; the other was registered to care for physically disabled adults under the age of 65. Our criterion for inclusion in the follow-up study was that the home was registered with CSCI or CSIW to provide long-term accommodation and care for older people. Twenty of the 37 eligible homes were revisited and some data were collected on the remaining 17 homes, as we explain below.

Our purpose was to compare the surviving homes then with now. For this reason, in the 20 homes we revisited, we replicated Townsend's method, but with some modification in order to accommodate not only historical and cultural changes in policy and practice but also changes in methods of social inquiry. There was no exact precedent for our method. As Charles et al. (2008, p. 40) point out,

> [r]estudies are relatively new in sociological research, in part because they raise methodological questions about the validity of comparison given the inevitable changes in both social and analytical contexts during the period between the original and the restudy.

In revisiting three classic community studies (Sheldon, 1948; Townsend, 1957; Willmott and Young, 1960) carried out some 50 years ago, Phillipson et al. (1998, p. 264) argue that simply replicating the previous research would not have been viable. One reason they put forward is that 'social science methods have moved on and different approaches are now available to explore the main research questions'. They therefore focused on some key findings from the previous studies and, using their own research design, set out to explore the extent to which these findings still held true in late 1990s. What is distinctive about our research is that we were following up a specific cohort of homes and this, in our view, required some replication of Townsend's methods although not his overall research design. The approach we adopted therefore was in line with that of Davies and Charles (2002, p. 13): 'the research design is replicated where meaningful and altered after careful consideration of social analytical changes where necessary'. They argued that such an approach would 'increase ... understanding of both the original object of study and the sociological account of it, as well as the contemporary object of the re-study and the intervening processes by which one was transformed into the other' (ibid.). Following this, we made some modifications to the content of Townsend's research schedules in order to take account of changes in the everyday life of older people since the 1950s.

The context of change

As we pointed out in Chapter 1, British society is in many ways barely recognizable when compared with the 1950s. Highlighting such change, Kynaston describes 'Austerity Britain' of the late 1940s and early 1950s as a world of

> [n]o supermarkets, no motorways, no teabags, no sliced bread, no frozen food, no flavoured crisps, no lager, no microwaves, no dishwashers, no Formica, no vinyl, no CDs, no computers, no mobiles, no duvets, no Pill, no trainers, no hoodies, no Starbucks.
>
> (2007, p. 20)

Of course within the two worlds of the 1950s and early 2000s there are complex variations related to age, cohort, gender, income and class affecting the conditions and experience of old age. For some, the 1950s is remembered as a time of post-war hardship, poverty and continuing rationing. But, increasing numbers of older people were 'neither impoverished nor unhappy' (Thane, 2000, p. 386). As the country slowly revived after the war, newly invented gadgets began to appear in homes as they became affordable: Dansette record players, television sets, food mixers, fridges and Formica surfaces. Supermarkets and self-help stores changed the experience of shopping dramatically, though such changes were not always understood or welcomed by older people (Nell et al., 2009). Despite evidence from Willmott and Young (1960) that there was a new role for older people within families in this time of growing consumerism, homemaking, fashion, cookery and childcare, Phillipson et al. suggest that 'with the benefit of hindsight, the 1950s must have been a difficult period in which to grow old ... older people were marginalised rather than liberated by greater affluence' (2001, p.15).

Although the creation of a welfare state altered the expectations of older people in regard to retirement, social protection and health care, there is evidence that the lives of a large number of older people in the 1950s remained difficult. Many lived in appalling conditions, in what Phillipson et al. have called 'the dilapidated urban environment' of bombed and damaged buildings where substantial proportions of households were sharing a bath, lavatory and kitchen sink with another family (2001, p. 6). Some lived alone in a single room with a curtained sleeping area, a sink in the corner and a shared outside toilet. It would be hardly surprising if some saw old people's homes as offering a more attractive alternative.

By comparison with the immediate post-war period, there has in general been greater affluence among older people throughout the 1990s and into the twenty-first century, enabling access to an ever-increasing range of consumer goods (Rees Jones et al., 2008). Advances in technology have also changed the lives of many, for example, by making it easier for older people to remain in their own homes or in sheltered accommodation with the help of a variety of assistive technologies (Peace et al., 2006). Whereas home phones were a rarity in the 1950s, now they are commonplace and those older people who require them now have access to specially adapted phones if they are visually or hearing impaired. In addition, in England in 2002, around three in ten men and nearly one in five women aged 80 and over said that they owned a mobile phone, while one in ten men and just over one in 20 women aged 80 and over reported using the Internet (Soule et al., 2005). This is a major difference in lifestyle from that in the 1950s, and these resources offer unprecedented opportunities for relationships and social networking among older people across distances regardless of domestic circumstances.

Accompanying the rise of the consumer society have been other significant social trends over the last 50 years which have important implications for intergenerational relationships and exchanges and the part that residential care might play in later life. Changes in household composition, for example, mean that older people are now much more likely to live alone or only with a spouse. Between 1961 and 2001, the proportion of people of pensionable age living alone doubled (Victor et al., 2009). However, although older people who live alone are more likely to feel lonely, Victor et al. found no statistical association between the two when all other factors are taken into account (2009, p. 209–10). Other changes include the increase in divorce and remarriage and the consequent reconstitution of families and family life, the increased participation of women in the labour market, the decline in a sense of solidarity due to the break-up of working-class communities and increased geographical mobility, the expanded middle class and the increase in inequality; and the rise in multiculturalism and racial diversity (Coates, 2005). All these trends have an impact on how older people can both give support to and receive support from family, friends and neighbours.

It was in the light of these kinds of changes, together with the substantial developments in research ethics (to which we return below), that we approached the follow-up study and the design of our research instruments.

Research instruments and methods

Before visiting the 173 homes, Townsend collected some basic demographic data on the residents of each home including information on sex, age, marital status, occupation, mobility, surviving children, contacts with visitors and length of stay. He used the latter information to select 'new' residents to interview. While visiting each home, he interviewed the matron or warden using a questionnaire (1962, p. 13). This included questions about the staff and residents, professional services, occupations and recreations, rules and routines, illness and death, administration and relationships with the local authority. He also completed a schedule detailing the physical amenities in the home. On the basis of this information, a report was written on each home. He used a further questionnaire to interview new residents (1962, pp. 457–62). In addition, a few diaries were kept for a week by residents and staff and Townsend took photographs.

The very process of redesigning Townsend's research schedules was indicative of just how much change there had been since the 1950s to the physical environment of care homes and the technologies available to them. Table 3.4 lists the data we collected and the means by which they were collected.

The Advance Information Schedules were based on the information collected by Townsend. For residents, it included age, sex, marital status, length of residence, frequency of visits from friends or relatives, whether they could leave the home unattended and whether they did, and whether they required help with dressing. We added a column on ethnicity, something that gets no mention in *The Last Refuge*, and we changed marital status to partnership status. For staff, in addition to staff category, age, sex and length of service, the schedule included information on working hours and pay rates. Again we added a column on ethnicity and another for agency workers.

The schedule used to interview new residents was modelled on Townsend's interview schedule (1962, pp. 457–63). This had to be modified in light of the kinds of family changes we have noted above, allowing for the possibility of step children or step grandchildren and taking into account new forms of contact and communication such

Table 3.4   Sources of data for the follow-up study, 2005–6

Research instruments/methods	Data collected
Advance Information Schedule (Residents)	Basic demographic data on 593 residents
Advance Information Schedule (Staff)	Basic demographic data on 659 staff
Home Manager Interview Schedule	Transcribed interviews with 20 managers
Buildings and Facilities Schedule	Physical environment and amenities in the 20 homes visited
Resident Interview Schedule (including the 'Incapacity for self-care' measure)	Transcribed interviews with 76 residents (including an 'incapacity' score)
One Week Diary	18 completed diaries (nine staff and nine residents)
Digital Photography	Photographs of the homes and the residents
Supplementary Material	Home brochures, menus, complaints procedures, etc.
1959 Quality Measure	Historical quality scores for 20 homes visited
CSCI and CSIW Inspection Reports	Modern quality score on 37 homes (approximately four per home)
Home Report	Report on each of the 20 homes visited

as mobile phones and the Internet, DVD players and recorders and personal television sets, and indeed the possibility that some residents might have a car or other form of transport.

The schedule used to interview the manager or deputy manager of the home about how the home was managed and run was a modified and considerably expanded version of Townsend's original questionnaire. Issues that barely featured in the late 1950s had to be added, for example, issues related to staff training and working conditions, to working with people with dementia or other mental health issues and to the administrative work relating to the legislative developments since the late 1950s, such as those regarding regulation. In consequence, the interviews we undertook with the home managers were lengthy and in some instances had to be broken up into two or three sessions. Sometimes the interviews were interrupted by unplanned events or crises, including the arrival in one instance of a CSCI inspector making an unannounced inspection visit. Such interruptions were of course an indication of what comprised the job of being a home manager in 2005–6, something we return to later in this book.

The present-day researcher may well argue that interviewing home managers is not a good way to find out what is going on in a care home, and therefore may question the value of adopting this approach. This is undoubtedly a valid argument and raises questions about the nature of truth. Our approach to this issue is an interpretive one, namely that our aim was to compare how wardens or matrons talked about the home in the late 1950s with how the current manager talks about it now. Neither may be representative of an objective truth, but they tell us about current (both then and now) values, opinions, prejudices, concerns and aspirations, and how these have changed or persisted over time.

Apart from the interviews, we, like Townsend and his colleagues, requested and were taken on a tour of the home. This together with short periods of observation enabled us to note its design, facilities and equipment. The 1958–9 schedule included such items as the number of bedside mats, wardrobes and chests of drawers in a room. Townsend's concern, as we explain in Chapter 8, was to calculate the ratio of these items to individuals: how many were sharing wardrobes, toilets, wash hand basins and chests of drawers, for example. Unsurprisingly, as we explain more fully in later chapters, we had to make some considerable adjustments to how we recorded the physical environment. In addition to using a modified schedule, we took a great many photographs which we could compare with the hundred or so photographs in the Townsend archive at Essex.

Our photographs enriched our other data as well as producing new insights once we subjected them to closer analysis. As Jordanova has argued, each kind of data source 'affords it own insight' (Jordanova, 2000, p. 191). Although photographs like other sources of data do not provide 'unbiased, objective documentation of the social and material world ... they can show characteristic attributes of people, objects and events that often elude even the most skillful wordsmiths' (Prosser and Schwartz, 1998, p. 116). Rose has written about the way in which photographs are particularly good at capturing the 'texture' of places and the 'elusive qualities that define sense of space' (2007, p. 147).

The periods of observation also enabled us to learn something about the daily life of staff and residents in the public spaces of the homes. The increase in privacy afforded to residents through the replacement of dormitories by single rooms and a greater consciousness of the need for privacy meant that a researcher who was only in a home for a few days would get a very partial view of home life. For example, apart from the moving and handling of residents in public spaces in the homes, we did not observe anybody work, one of the key tasks of care workers and one through which power and control can be exercised and indeed abused (Lee-Treweek, 1994; Twigg, 2006). Nor did we witness any private exchanges, between or among staff and residents. And, like Townsend's, our research did not include being in the home during the night shift. However, we did obtain insight into the private lives of residents and staff through the 18 diaries we received: nine from staff and nine from residents.

Following our visits to a home, we produced a detailed report based on an analysis of the completed schedules and interviews outlined in Table 3.4. Very few of Townsend's completed schedules have been preserved but, by producing a home report which replicated the format of Townsend's home reports, we were able to make some detailed comparisons of the homes then and now.

In addition to the home reports, two quality measures for each home were completed. The first used Townsend's original 40-item measure (Townsend, 1962, Appendix 3). The second was a modern quality measure derived from a series of CSCI inspection reports on each home. We explain this aspect of our methodology in more detail in Chapter 8. Using these measures we were able to compare the quality of the homes not only with each other in 2006 but also across time. Using the relevant CSCI inspection reports, we were able to give the 17 homes which were not visited a modern quality rating and to compare these reports with Townsend's home reports.

All these procedures generated detailed comparative data relating to the built environment, staffing and routines, the characteristics of residents and reasons for admission, and their daily lives. This allowed us to explore not only continuity and change in the care regimes, in the quality of care provision and in the conduct of social research but also in cultural constructions of, and responses to, old age (see, for example, Rolph et al., 2009).

Access, participation and consent

In 1959 the issues of obtaining access, securing participation and consent were rather different. At this time social research was not subjected to formal ethical clearance procedures. Our research, as with all in this field in the early 2000s, was circumscribed by a variety of research governance and ethical procedures including having the proposal scrutinized and approved by an ethics committee and obtaining Criminal Record Bureau clearance certificates for ourselves. During the course of our research, of the 37 homes eligible to be included in our follow-up study, 29 were contacted by us and invited to participate. Two, despite numerous requests never responded; both were small private homes. A further two refused to participate on the grounds of shortage of time: one was a small voluntary home; the other was a small private home. The remaining 25 homes readily agreed to participate and the majority showed a keen interest in the historical dimension to our research.

The managers of the 20 homes selected for inclusion in the study were sent a package of leaflets about the project to distribute to every resident and member of staff. We wanted to ensure that as far as possible everyone in the home would be involved in the decision to participate and would not be left wondering when we visited, who we were and what we were doing. The introductory leaflet included an invitation for any member of staff or any resident to keep a diary for a week. One reason for this was that Townsend also elicited some diaries from residents and staff (1962, p. 14). But a second reason was that we wanted to offer everyone the opportunity to participate in some way and not just the few selected for interview. In following Townsend's method, we would only be interviewing the manager of the home and some newly admitted residents.

Having obtained consent and answered any questions, we then sent the home manager two 'Advance Information Schedules' to complete, providing us with basic demographic data on all staff and residents. This included information on length of residence, and our intention was to identify four residents who had been living in the home for a period of 6–12 months and to approach them by letter seeking their agreement to be interviewed. Such a procedure would not have been adopted in the 1950s. Rather, Townsend and his colleagues had to rely upon identifying suitable interviewees once they had arrived in the home. Inevitably, our procedures did not always go according to plan. Although a majority of the 20 homes we visited did complete and return the schedules in advance, two only completed them when we arrived, a few only partially completed them and a further two, small private home businesses, did not complete them at all. In these instances, procedures regarding whom we might interview were to an extent compromised. Rather than the resident having sufficient time to read the information about the project and what participation would involve, and being able to discuss it with a friend or relative if they wished, they had to make up their mind during the course of our two- or three-day visit. Conversely, however, a few residents had changed their minds since agreeing to be interviewed or felt indisposed, and a replacement therefore had to be

negotiated. By the time we had completed our fieldwork, we had interviewed 75 residents in the 20 homes.

The use of consent forms was a further procedure that would not
have been typical in the 1950s. Our consent forms, in line with current
guidance at the time (see, for example, ESDS, nda, ndb; Social Research
Association, 2003; Department of Health, 2005b) were designed to
serve a variety of purposes. First, they were used to ensure that those
we interviewed had been fully informed about the research and had
the opportunity to ask questions about it. Second, they were used to
obtain permissions, not only to interview but also to photograph and
to archive data for use by future researchers. Striking the right balance
between putting people at ease and at the same time ensuring that such
procedures are being fully observed is not straightforward, particularly
with people who have some memory difficulties (Hubbard et al., 2002).
We did not wish to exclude people with dementia and had to rely on
the view of the manager as to whether the residents we had selected
were capable of giving informed consent. Our overall approach to
consent, however, particularly in regard to photography, was to treat
it as an iterative process, with ongoing permissions being ascertained
at different stages of the research, with attention being paid to different levels of consent, enabling choices for different uses and audiences
(Lewis, 2004).

The consent procedures had a particular impact on the kinds of
photographs we were able to take or replicate. Some of Townsend's
photographs were of groups of residents and although such scenes have
their modern-day equivalent, it would have been impossible to obtain
consent from all participants. Other photographs taken by Townsend
might today be considered too intrusive. Consequently, our own photographs portrayed a different reality and one that represents older people
in a different way when compared with 50 years ago. We return to this
issue in Chapter 9.

Methodological issues

Our design raises many methodological issues, some of which we have
discussed elsewhere (see, for example, Johnson et al., 2007), and some
of which are discussed in more detail in Chapter 9 of this book. First,
there are ethical and quality issues related to the use of volunteer or
'lay' researchers. Was our relationship with our volunteers exploita-
tive? How reliable were the data they provided? Suffice it to say at this
point, that the quality and quantity of information generated by the

volunteer researchers was, overall, very high (see Appendix 2). This was, in the main, due to the kinds of people who responded to our call for recruits. Some were experienced local historians, and, as mentioned earlier, 60 per cent had a first or higher university degree. It was also due to careful preparation on our part. Before funding was sought, the tracing study was thoroughly piloted by us and by one volunteer. The volunteers were also comprehensively briefed and supported, both individually and through our newsletters (www.lastrefugerevisited.org.uk).

A second set of issues relates to changes in social research methods and procedures to which we have already alluded. Townsend's research was undertaken in the positivist tradition where the distinction between the 'known and the knower' (Lincoln and Guba, 1985) was retained, and the method of inquiry was regarded as value free. Although we have replicated his method, 50 years on, we have taken a more reflexive approach not only to the origins of the primary data but also to the production of our own data. Furthermore, ethical procedures, such as those governing consent, are indicative of the agency now ascribed to older people in care homes in contrast to 50 years ago when they were seen as passive victims of the institutional system (Phillipson et al., 2001). It may also be argued, however, that they are indicative of the overprotection of those categorized as vulnerable and of the defensive practices of those charged with their care.

A third set of issues relates to archiving. Both Townsend's data and our own have been deposited in the UKDA. This raises further issues not only about the ethics of reuse and secondary analysis but also about consent, anonymization, confidentiality and access. Our research combines both sociological and historical methods and these may sometimes conflict. A balance has to be struck between guaranteeing the anonymity of one's sources of evidence and acknowledging personal testimonies. Furthermore, longitudinal research requires the preservation of identity for future research. In archiving our data we have had to confront the difficulties these issues pose.

Part II Revisiting *The Last Refuge*

4
Survivors and Non-Survivors

It was through the tracing study that we were able to identify surviving and non-surviving homes. The reports completed by the volunteers on the homes that had closed and by us on those that had survived, provide information on the history of each home since 1959. Drawing on these data we address, in this chapter, the key issues of what type of homes closed between 1958–9 and 2005–6, when they closed and why. The data reveal a pattern of attrition that can be related to changes in social and public policy. They also indicate what kinds of usage surviving buildings have been put to and what changes in ownership have occurred. For example, some have been demolished and replaced by new buildings containing new services for older people such as age-restricted housing. Other buildings have been converted for different uses such as the two that are now museums. One is a doctor's surgery and another a veterinary practice. Some are now day centres, others hotels, and some are once again private residences. We also look at the 39 homes that survived as care homes and consider some of the factors that might account for their survival.

As mentioned in the previous chapter, in *The Last Refuge* Townsend devotes a chapter to each of the four types of homes he visited: the 39 ex-PAIs, the 53 other local authority homes, the 39 voluntary homes and the 42 private homes. We have organized this chapter to reflect this and to draw out some of the differences between the histories of the homes in these sectors. First we examine the pattern of closures in each sector. Then we look at the surviving homes and draw some comparisons with the non-survivors.

Figure 4.1 shows the pattern of closures of the 134 homes that have not survived. The highest attrition rates are in the public sector homes.

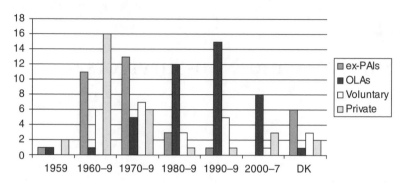

Figure 4.1   Home closures by original tenure, 1959 to 2007

The ex-PAIs

The story of what happened to the 39 ex-PAIs, their residents and the sites upon which they stood is particularly complicated. This is because of their variety and complexity. Many in 1959 were part of extremely large complexes serving a variety of functions. The 17 joint-user institutions included in Townsend's sample, for instance, provided NHS hospital care as well as residential care which was the responsibility of the local authority welfare committee. Several of the ex-PAIs still accommodated a range of residents, not just older people, and a high proportion (47 per cent), when compared with other tenures and with today's gender profile of care home residents, were men. Some of these institutions were enormous. Luxborough Lodge, for example, on the site of which now stands the main campus of the University of Westminster, accommodated 1250 people. At the other extreme, Burlington House in Yorkshire accommodated only 65. Some were set in large acreages of land such as St Anne's in Surrey, which occupied 25 acres, while others occupied fewer than two acres.

Townsend was particularly critical of conditions in these ex-PAIs and he used his photographs to illustrate the 'scandalously low' standards of care and accommodation that many of them provided, even those where some attempt at improvement had been made. In his discussion of future policy, he drew attention to the 'large-scale programme of modernization and improvement' that would have to be undertaken if the old workhouses were not to be demolished (1962, p. 417). He argued that the costs involved could well match or exceed the cost of replacements and that the latter would be a much better long-term investment. He recommended, therefore, that they should be closed and that 'many of the buildings should be demolished and the sites sold or used for other purposes' (ibid., p. 418). As Figure 4.1 indicates, their closures began soon after he completed his research, peaking in the late 1960s and early 1970s. This coincides with the expansion of the building programme of new, smaller residential care homes. In 1970 alone, 100 new homes were built accommodating 4544 residents and 14 PAIs were closed (Peace et al., 1997, p. 12).

By 1993 all but four of the 39 ex-PAIs Townsend visited had ceased to provide on-site residential care facilities for older people and the residents had been dispersed to other homes. Westbury Hall, for example, closed as a residential care home in 1969. It was a joint unit and in addition to providing residential care, the building was also used by the hospital management committee to accommodate 40 'geriatric' patients. By 1968, all the patients had been moved to a new geriatric unit six miles away. At the same time three new residential care homes had been built – one a quarter-of-a-mile away and the other two in nearby towns. Of the 44 residents remaining in 1969, 18 were transferred to the nearest (35-bedded) home and the remainder to other county council homes. The old workhouse was demolished shortly afterwards. In 1973 the site was sold by the County Council to the Rural District Council and council housing was built on it: 12 two-bedroomed flats, a terrace of five three-bedroomed houses and a terrace of three old people's bungalows. Half of this accommodation is now owner occupied and in 2003, the other half was transferred to a registered social landlord. In 1989, part of the site was sold by the Rural District Council to a private developer and this is now occupied by 25 three- and four-bedroomed houses.

Like Westbury Hall, nearly two-thirds of the 39 ex-PAIs have been completely demolished. Regarding the rest, the land and those buildings that are still standing have been used, in the main, in the public interest, though the link to older people with care and support needs is relatively weak. Several of the remaining buildings have preservation orders on them, such as the clock tower of the hospital at The Poplars, Wolverhampton, and the former Union Workhouse (now a museum) at Gressenhall, Norfolk.

Of the four ex-PAIs where there is still a functioning care home on site, all were purpose-built to replace the old buildings. Two were built in the 1960s and two in the 1970s and, with the exception of one (Bridge House), the old buildings were then demolished. Those belonging to Bridge House were sold by the county council in 1978 and are now part of a farm and museum. Ownership of two of these existing homes was transferred to the voluntary sector in the late 1990s, in both cases as part of the bulk transfer of care homes by their respective local authorities to large not-for-profit companies. The other two have remained in public ownership.

Other local authority homes

Only five of the 53 'other local authority homes' visited by Townsend were purpose-built in the post-war period; the remainder were mainly Georgian and Victorian houses converted for use as residential care homes in the 1940s. Many were listed buildings. Townsend describes their former opulence as somewhat forbidding and was critical of the remoteness of some, being situated down long drives and distant from urban centres, shops and local transport. He also comments on 'awkward passages and stairwells', the shortage of and lack of accessibility to toilets and the low proportion of single bedrooms and ground floor accommodation. Even in these homes, residents did not always have exclusive access to such items of furniture as wardrobes, chests of drawers, washbasins or dressing tables, and furniture was sometimes 'shoddy' and had been second-hand at the time of purchase (1962, pp. 115–8).

Townsend recommended that these homes, like the ex-PAIs, should be closed with priority being given to the closure of remotely sited homes, larger homes with 40 or more beds, those with few or no single rooms and those with poor physical facilities. He further suggested that half the population accommodated in these homes and the ex-PAIs should be accommodated in replacement premises for long-term and short-term nursing care and provided very detailed specifications on how these should be sited, equipped and staffed, and the principles upon which they should be managed.

Figure 4.1 shows that closures of the other local authority homes he visited started some ten years later than the ex-PAIs, peaking in the 1980s and 1990s when, as we discussed in Chapter 2, the role of local authorities as the major provider of residential care started to decline dramatically. By December 2006, 43 had closed and, with the exception of three, all had remained in local authority ownership until closure. Fairhaven (see Appendix 2) is a typical example of the closures in the early 1990s when implementation of the National Health Service and Community Care Act 1990 began. This was the point when local authority homes would for the first time be subject to inspection procedures alongside the private and voluntary sector homes. As a Grade II listed building, the costs of upgrading Fairhaven were prohibitive. This, combined with the shift towards community care and the costs of implementing the Children Act 1989, led to the decision by the county council to close Fairhaven and some of its other residential care homes. Like Fairhaven, several of the homes that closed around this time were listed buildings which were no longer fit for purpose and deemed 'surplus to requirements'. Although many, following closure, were sold to private developers, several remained in local authority ownership and were converted for alternative uses including resource centres for children or people with learning difficulties and a day centre for

Photograph 4.1 Four former homes now closed: An ex-PAI under restoration, 2005 (*top left*); a small private home now a private residence (*top right*); a voluntary home now a hostel (*bottom right*); a converted local authority home now special housing (*bottom left*)
Source: Barbara Prynn, Peter Wade (top right). © Barbara Prynn (top left, bottom left and bottom right), © Peter Wade (top right)

older people. Some were transferred to registered social landlords and developed into sheltered or extra care housing schemes.

Of the ten that survived, only two (both in Wales) are still owned by the local authority, six are now in the private sector and two in the voluntary sector. The transfer of ownership from the public to the 'independent' sector reflects the impact of the National Health Service and Community Act 1990 and the Care Standards Act 2000, both of which steered local authorities towards divesting themselves of their care homes.

The voluntary homes

Townsend notes that homes run by voluntary organizations were looked upon favourably by the Nuffield Survey Committee which published its report on the care of older people a year before the passing of the National Assistance Act 1948 (Rowntree, 1947). Hence their numbers almost doubled between 1948 and 1954, after which growth in this sector fell off sharply (Townsend, 1962, pp. 151–2). The 39 voluntary homes visited in 1958–9 were, broadly speaking, of two types. First there were those (15) that were run by faith-based organizations ranging from a variety of Roman Catholic and Anglican Orders to organizations such as the Salvation Army and Methodist Homes for the Aged. Second were those (24) run by secular organizations and small charitable trusts. These included a wide variety of organizations serving the needs of specific communities of identity or locality. Townsend reported that about half of the voluntary homes were very remote from shops and urban centres. Although some provided 'a large measure of comfort and privacy', others were 'bleak in their amenities and furnishings' (1962, p. 176). When compared with the local authority homes, more of the residents were women, unmarried and middle class. Furthermore, although of similar age to those in the local authority homes, the residents of the voluntary homes were fitter and more active. Indeed, somewhat contravening the purpose of the National Assistance Act to provide for those 'in need of care and attention', it was a requirement of several that residents should be able to look after themselves. Nearly all the homes were run by women, a third of whom were aged 60 or more. Overall, Townsend found these homes to be more liberal in their regimes and provided more privacy. It was his view, however, that those that provided for more active people should, with the assistance of government, focus their efforts on the provision of sheltered housing rather than residential care.

Although a majority (25) of these voluntary homes have closed, unlike the public and private homes, closures in this sector have been fewer and more evenly spread over time with no marked peaks and troughs.

The faith-based homes

Ten that closed were run by faith-based organizations including three run by Anglican Orders (Society of Servants of Mary, Order of Community of St Mary, Anglican Sisters of Bethany), one by the Jewish Convalescent and Aged Homes, one by the Salvation Army, and five by Roman Catholic Orders (Little Sisters of the Poor, Poor Sisters of Nazareth, Congregation of Our Lady of Pity). Townsend's sample included four very large voluntary homes, all owned by Little Sisters of the Poor; three of these have closed and have been demolished.

Although one of the smaller Roman Catholic convents still provides accommodation for six retired priests, it is no longer registered as a care home. As Mother Superior explained, in 1959 when Townsend visited, there were four 'elderly ladies' being cared for in the convent. In 1963 the convent started to provide nursing home care for retired priests and by 1997, when it ceased to function as a care home, the nuns had cared for around 250 priests. Not all were 'sick' however. So, in the mid-1970s the stable block was converted into self-contained flats and three bungalows were built for use by those who wanted to be more autonomous. In the early 1990s, the new legislation regarding registration and inspection began to threaten the way of life of the convent community and its rationale. For example, the nuns had no wages or normal working hours, but the inspection unit wanted them to conform to these kinds of institutional practices. They had to comply with new fire regulations and to have thermostats fitted on all the taps. They had to fill up long questionnaires for the inspectors, most of which they considered to be not applicable to them. Their eventual solution was to change their function so that they would not be forced to compromise their way of life. They decided to offer men and women in religious communities who wanted a holiday or retreat, or even a sabbatical for six months or a year, the opportunity to rent one of their cottages or flats. Now they have visitors from as far away as the US and Australia. The six priests who live in the main house are retired 'but mobile'. These men, we were told, were not well enough to run a parish any more but continued to help out in the parish and to visit people to say mass.

A key reason why this convent community has survived is that it is extremely well funded. The bishop has raised very large sums of money which are ring-fenced for the care of retired priests in the diocese. It is an interesting example of how the Order has managed the conflict between its own mission and that of the CSCI. The other seven convents, however, closed due to lack of resources (in particular, it would seem, in the case of the Anglican homes) or as part of a more general rationalization plan. One of the Anglican homes, for example, was a large Jacobean mansion in 25 acres which was sold in 1960 and became a private school. Another of the Anglican homes closed because of a shortage of women wanting to join the Order; the remaining sisters moved to another convent which became the motherhouse. This pattern of closures is reflected by Philpot (2003a, 2003b) who reported on research on 238 religious orders in England, 50 of which were providing residential care as mission, as well as for their retired members. Some of the other 188 organizations provided such care exclusively for retired members. The main focus of Philpot's report was on the familiar problem of achieving required standards of care under the Care Standards Act 2000 and the low level of fees charged in relation to costs. Over half of the 50 orders had closed homes or were planning to do so and several more stated they were 'just surviving'. A total of 37 homes had closed in the decade up to 2002, and ten per cent of the residents had died during the closure and transfer process. Just over a third of the orders had developed new kinds of work following closure, including running retreat houses, hospital chaplaincy, parish catechetics, working with drug users, with prisoners and their families, counselling and teaching.

Homes run by secular organizations

The remaining 15 homes of the 25 that have closed include two run by regional associations for the blind, one by a local association for the deaf, a home owned by the Red Cross, with the remainder belonging, in the main, to locally based housing associations or old people's welfare committees such as the Manchester-based Gentlewomen's Housing Association, Cheltenham Old People's Housing Society and Birmingham Council for Old People. All were converted, mainly Victorian, houses. The vast majority of residents in these homes were women and half the homes accommodated only women, possibly in shared rooms with shared bathing and toilet facilities.

Most were closed because the costs of maintenance and refurbishment were prohibitive, and in some instances this was combined with the fact that the demand from a specific clientele for this kind of service had diminished. For example, an Association for the Blind sold its home to the county council in 1973 so that residents could move to locally based homes. The home itself then became a non-specialist home for sighted and non-sighted people until it closed in the mid-1980s. A further example is the home for 'gentlewomen from Devon' which belonged to an endowment trust. It closed in 1976 because there were insufficient numbers of gentlewomen in straightened circumstances requiring its services. It was sold, but ten years later reopened as a nursing home. In 2000 it was sold to a developer with plans for conversion to flats.

As with these two examples, nearly all the 25 charities owning these secular voluntary homes continue to exist but since the early 2000s most have diversified and now provide different kinds of support services to their clientele – much as Townsend had suggested. Five still own and manage a few care homes and/or sheltered housing and another still owns two blocks of almshouses. Only two no longer exist and a third, the Manchester-based Gentlewomen's Housing Association, transferred all its assets to another registered charity, the Northern Counties (Specialised) Housing Association.

The private homes

The private homes Townsend visited were distinctive in a number of ways when compared with the public and voluntary sector homes. For a start they were considerably smaller. Of the 42 he visited, just six accommodated 20 or more residents. The largest was a former hotel on the Hampshire coast which accommodated 38 people. Indeed, all but one of these larger homes had formerly been seaside hotels. This reflects the role played at the time by some hotels, guest houses and boarding houses in providing accommodation for older people with the means to pay. Not only were the private homes small when compared to some of the very large local authority and voluntary homes but also they were concentrated in particular parts of the country; 76 per cent of Townsend's sample of private homes were in London and the south of England. Many were in the southern coastal resorts to which the middle class often retired, away from family (Law and Warnes, 1973; Karn, 1977). In Wales and some parts of the north and midlands of England there were no private homes from which Townsend could draw a sample.

Also, unlike the public sector homes which housed almost equal numbers of men and women, the majority (four-fifths) of the residents in these private homes were (mainly middle class) women (Townsend, 1962, Table 16, p. 56). Two-fifths accommodated women only. Although half of the residents had no surviving children, fewer were 'unmarried or otherwise childless' than in the public and voluntary sector homes. Furthermore, the residents in the private homes were older: two-thirds of the women (compared to half of the men) were over 80 and a fifth were in their 90s. There was also a higher proportion of 'extremely frail' residents than in the other types of homes.

Almost all the homes (39 of the 42) were family-run businesses which were managed by the proprietors providing a home not only for the residents but also for themselves. Nearly two-thirds of these proprietors were women, most of whom were widowed, separated or divorced. A third of them were in their 70s and less than half had any kind of training. They were supported by other family members and what Townsend describes as an 'odd mixture' of people who 'were often content with a small wage', such as

> the male friend who took a month off in London when he felt like it; the woman who had been a children's nanny; the woman whose face was plastered with rouge and lipstick and who was studded with dress jewellery; the Polish refugee from a Nazi concentration camp; mentally subnormal girls on licence from local hospitals and young girls from Spain and Germany.
>
> (Townsend, 1962, p. 187)

Fees in these homes ranged from three-and-a-half to 30 guineas per week. The quality of care was likewise extremely varied ranging, for example, from the former hotel on the south coast of England where all the residents had single rooms, to the end-of-terrace house where the residents were accommodated in tiny shared rooms, with few facilities, no heating, plywood doors, flaking paint and rotten floorboards (Townsend, 1962, pp. 180–1). These findings led Townsend to comment on the inadequacies of the system of registration and inspection under the National Assistance Act 1948.

Tracing the history of these homes when compared with the public and voluntary homes was certainly more difficult due to the scarcity of records. Nevertheless our volunteers managed to trace the history of most and, as Figure 4.1 indicates, just over three quarters of the closures (24) occurred between 1959 and 1979, peaking in the 1960s. In some instances, we know that the owners retired and closed down the business. For example, a house in north London was owned by a Polish couple who took in four residents. They retired in 1968 and the house reverted to being their family home which they shared with their married daughter who still lives there. They also owned a home around the corner which housed a further five residents. This ceased as a care home in 1964, when it was converted into two flats. One was sold. The other still belongs to the daughter and is rented out. Like this example, ten other owners remained in the home after it was closed as a residential care home and continued to live in it as a private dwelling or as another business, such as a guest house or a hotel. The remainder included two where the owners died, seemingly while the homes were still functioning as care homes, and there is no information about what became of their businesses. The rest sold their properties and moved elsewhere. The new owners converted the premises for a variety of uses: flats, hotels, guesthouses, business offices and student accommodation.

During the next 20 years (between 1980 and 1999), there were only two closures in these private sector homes. This coincides with the period during which the Department of Social Security (DSS) board and lodging payments for people in residential care became a right (1983) and the concomitant expansion in the provision of private residential care for older people. Of the two that closed, one closed before the boom began and the other after it was over. Both were in what had become 'run down' areas of high multiple occupation; one in London and the other on the south coast.

After 2000, closures began to rise again, three over a period of six years. One turned into a 'home care' service, using the former home as its offices; the other two were sold and converted to private dwellings. These closures reflect the impact of

(a) the NHS and Community Care Act 1990 which ended the availability of DSS payments, transferring financial responsibility back to local authorities, thereby limiting funding streams for people wanting to enter residential care;
(b) the shift away from residential care towards care in the community thereby reducing demand;
(c) the anticipated consequences of the Care Standards Act 2000 which required homes to meet certain minimum standards which many homes claimed they could not afford to meet, thus putting them out of business and
(d) the introduction of the minimum wage which interacted with price controls effected through local authority commissioning policies.

As pointed out in Chapter 2, the reasons for the rise in closures from the late 1990s onwards were the financial ramifications of these developments for small businesses.

To understand more fully why these homes closed and more particularly why they closed while others survived, we turn now to look at the surviving homes.

Accounting for survival

Table 4.1 shows that nearly one in four of the 173 homes Townsend visited survived. The category with the highest survival rate is the former voluntary homes, followed by the former private homes with the former local authority homes having the lowest survival rates. A number of interrelated factors may account for these differences.

Table 4.1   Survival rates by 1958–9 tenure

1958–9 tenure	1958–9 sample (n)	2005–6 survivors (n)	Survival rate (%)
Ex-PAIs	39	4	10.3
Other local authority	53	10	18.9
Voluntary	39	14	35.9
Private	42	11	26.2
Total	173	39	22.5

Tenure and ownership

Clearly tenure is a factor. Table 4.2 shows how the tenure of the surviving homes has changed over the years. In 1959, about a third of the 39 surviving homes belonged to the local authority, a third to the voluntary sector and a third to the private sector. In 2006, only ten per cent of these were still owned by local authorities with the remaining 90 per cent being equally distributed between the voluntary and private sectors. One way that homes have survived, therefore, has been through the transfer of local authority homes to the private sector and voluntary sectors, predominantly the former. Several now belong to

Table 4.2   Surviving homes by tenure, 1958–9 and 2005–6

Type of tenure	Tenure in 1958–9 (n)	Tenure in 2005–6 (n)
Local authority (inc. ex-PAIs)	14	4
Voluntary	14	18
Private	11	17
Total	39	39

large non-profit trusts or private for-profit companies. In April 1999, Hillcrest for example was transferred along with all the other residential care homes in the same county to a large non-profit organization. In 2006, this was the second largest non-profit care provider in the UK operating 75 homes in four different counties. Likewise, in 1997, Rivermead became part of a bulk transfer of homes to another non-profit company. In the early 2000s, Fairfax House was sold by the local authority to a large corporate provider which was later merged with one of the largest for-profit providers in the UK. These trends match national trends as discussed in Chapter 2 and reflect the impact of national policies on local provision.

What Table 4.2 also demonstrates is the relative stability of the former voluntary and private sector homes. Unlike the former local authority homes, few have changed tenure. If we look at ownership, however, we find some important differences between the voluntary and private sectors. Twelve of the 14 former voluntary homes are still owned and run by the same charity as in 1958–9. These include some large organizations such as Methodist Homes for the Aged, the Salvation Army and the Royal British Legion, as well as smaller, specialized charities and individual homes arising from a single bequest. These homes have a long and stable history. By way of example, Overton Court, a home serving a particular community of interest, showed us many documents (including yearbooks) and photographs which recorded its history from its founding in 1934. A few of the residents had lived there for over 20 years. Like this home, several of the voluntary homes, we found, have photograph archives illustrating their histories. It is perhaps this sense of history that has contributed to what Kellaher (2000) in her study of homes run by Methodist Homes for the Aged, referred to as 'mutuality'. She argued that recruitment through the Methodist network tended to lead to the arrival of people who could become 'kindred spirits' (2000, p. 85), while retaining their individuality and personal integrity. However, she expressed concern about the future potential for mutuality as a result of the recruitment of increased numbers of frail residents.

In contrast, nine of the former private homes, while remaining in the private market, have changed owners at least twice although, remarkably, five are still small family-owned businesses. Monmouth House, for example, was sold to the matron of the home who then cared for the owner until she died. She sold it to the present owners in 1982 and it is now managed by their daughter. A further three have been turned into small limited companies. Only one, Pine Grange, has gone into corporate ownership and has been the subject of several takeovers in recent years involving numerous changes of manager. This home was opened as a private home in 1952 by a missionary returning from China who, with his family of six, had been interned by the Japanese army during the Second World War. One of the residents was his wife's uncle and the home was run very much like a family. It was sold to new family owners in the 1960s and again in the 1970s and in the 1980s when the new owner set up his own company eventually acquiring 25 homes. In 1997 his company went into receivership and was then bought by Highclear Homes. In 2004, it was bought by Ashbourne and in 2005 by Southern Cross, Britain's largest care homes operator, bringing dramatic changes to the way the home was run. The subsequent fortunes of Southern Cross suggest that change and instability have continued.[1]

With this one exception, the ownership of these former private homes has remained remarkably stable suggesting that they are a distinctive subset of homes which is well characterized by the term 'cottage industry' (Johnson, 1983; Peace and Holland, 2001) that has functioned in ways that are modelled upon ordinary family life. Survival, as with ordinary family homes, depends upon the business either being inherited by the younger generation, or being sold to someone else who is also prepared to take on the task of managing a small-scale business.

However, with the corporate providers increasing their share of the market, these small businesses are in a vulnerable position, and as the example of Pine Grange above indicates, homes in this sector may have more volatile trajectories with frequent changes of management and staff. The eight homes which became part of the private market in later decades, most of which, as mentioned above, were sold by local authorities to the private sector, illustrate this point. In 2006, for example, Highfields, a former local authority home now corporately owned, had been without a registered manager since 2003. The high turnover of managers in private sector homes when compared with the voluntary sector was commented on by Darton and Wright who found that just six per cent of home managers in the private sector homes they surveyed had been in post for 15 years or more compared with 63 per cent in the voluntary homes (1990, p. 84). Overall, therefore, the current 17 private homes among the 39 survivors were, in 2005–6, characterized by far less stability than the 18 voluntary sector homes.

Location

A further factor relating to survival is geographical location (see Figure 4.2) and there are some significant differences in this respect.

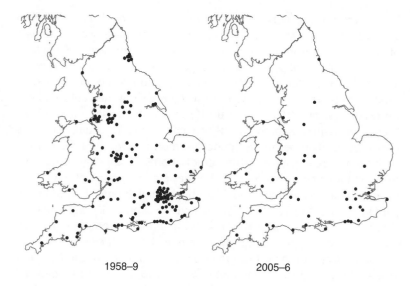

1958–9   2005–6

Figure 4.2 The location of the sampled homes, 1958–9, and the surviving homes, 2005–6

As Table 4.3 shows, the highest survival rates are in Wales and the south-west of England, and the lowest in the north of England. However, whereas the private homes and ex-PAIs only feature in those areas with the higher survival rates, the surviving former voluntary

Table 4.3 Survival rates by tenure and location

Location	Survival rate (%)	2007 survivors by 1958–9 tenure				
		Ex-PAIs (n)	OLAs (n)	Vol (n)	Priv (n)	Total (n)
North-West	4.2	—	—	1	—	1
Yorks and Humberside	7.7	—	1	—	—	1
Greater London	10.7	—	1	2	—	3
East Anglia	11.1	—	—	1	—	1
North-East	16.6	—	—	1	—	1
East Midlands	20.0	—	—	1	—	1
South-East	25.6	—	2	4	5	11
West Midlands	28.6	1	—	1	2	4
South-West	45.8	2	3	2	4	11
Wales	71.4	1	3*	1	—	5
Total	22.5	4	10	14	11	39

Note: * Although located in Wales, one of these three homes belonged to an English county council in 1958–9.

homes are more evenly spread across the country. This may reflect the impact of market forces on the private for-profit sector: survival of the former private homes is in the more affluent areas of England where demand from self-funders is higher. These are areas where there are the highest rates of owner occupation among older people whose capital assets are taken into account when assessing their ability to pay for residential care (Hancock et al., 1999). Although the former private homes we visited accommodated residents who were financially supported by local authorities, evidence from other research indicates that such homes are reliant on 'cross subsidisation' (Wright, 2003), meaning that the lower rates that local authorities are willing or able to pay to meet the charges are compensated for by the higher rates charged to and paid by self-funding residents for the same level of service. This suggests that self-funders are crucial to the survival of private homes in the small business sector. At the very least there must be a sufficient proportion of self-funders for the business to be sustained and a profit to be made. Five of the 11 former private homes are in seaside locations which are popular retirement areas (Law and Warnes, 1973; Karn, 1977) and where there is demand from their relatively high older populations who do not have family close at hand. The fortunes of the private homes which are now part of the corporate market are affected by different financial mechanisms: profits, as the example of Pine Grange demonstrates, are made primarily through trading (Drakeford, 2006; Scourfield, 2007).

Of the three surviving homes in the north of England, two are former voluntary homes and remain so and a third is a former local authority home which was sold as a small business. What is also striking in regard to location is that in Wales not only have most of the few homes visited by Townsend survived but three of the five survivors belonged to Welsh local authorities and still do. In Wales, a traditional Labour stronghold from which several architects of the welfare state emerged, commitment to public ownership has been stronger. In addition, public sector care homes in Wales are obliged to comply with the requirements of their counties' language schemes. Data from a study undertaken on the implementation of Welsh language policies in care homes in Wales found that a higher percentage of Welsh speakers lived in the public sector homes (62 per cent of the sample) than in the voluntary and private sector homes (Cwmni Iaith, 2002). This suggests that the public sector homes in Wales are meeting the demands of a particular market and our fieldwork in three Welsh homes confirms this. Two of the three were run by large national voluntary organizations and, in both, the proportion of residents who were Welsh speakers was very small, and neither employed any Welsh-speaking staff. In stark contrast, in the third home, owned by the local authority, Welsh was clearly the first language and was freely spoken during the course of our fieldwork by staff and residents alike.

Extensions, reductions and improvements

The number of occupied beds in the 20 surviving homes we visited ranged from eight to 203 in 1958–9, whereas the comparable range in 2005–6 was from 14 to 58. Table 4.4 shows that the general trend has been for the smaller homes, those with under 20 places in 1958–9, to increase in size and the very large homes, with over a hundred beds, to decrease in size.

The private homes, when compared with all the homes visited by Townsend, were much smaller, ranging from six to only 26. One might have expected the larger of these homes to be more likely to survive (Darton, 2004). However, this was not necessarily the case; the very small private homes were just as likely to survive as the larger ones. Of those with fewer than 20 residents in 1958–9, all were private family businesses. These are the homes that, over the years, have had to increase in size in order to remain viable. Between 1958–9 and 2005–6 their median size has more than doubled from ten to 26. This of course has involved substantial building works and improvements. One, for example, which accommodated eight residents in 1958–9, was a two-storey Victorian house which was first extended in 1994, when four bedrooms were added, and again in 2000, when a two-storey extension was added, so that when we visited it, it accommodated 19 residents. Another, having increased in size from 15 to 23 by building a small flat-roofed extension in the 1960s and converting former staff accommodation, was in 2006 preparing to embark on building a £1 million extension to increase its capacity from 23 to 39 places

Table 4.4 The size of the surviving homes in 1958–9 by their size in 2005–6

Size in 2005–6	Size in 1958–9					
	Under 20 (n)	20–9 (n)	30–9 (n)	40–9 (n)	50–99 (n)	100 or more (n)
Smaller	0	0	1	1	1	6
About the same*	2	7	1	1	1	0
Larger	7	2	0	2	1	0
Totals	9	9	2	4	3	6

Note: * A difference of no more than five places.

(including 11 independent living units). The proprietor told us that this was the only way his business could survive.

> [O]ur expansion plan has to go ahead because it's our only route of survival. As good as we are, we won't survive – we're too small now. There's very few independent homes left now. We're buoyant you know, we're fairly OK but as profits start to dwindle we can only put our fees up so much but costs keep spiralling. So once we have another 16 beds that should put us in a much more buoyant position.
>
> (MI/18)

Interestingly, his home was in a Grade II listed building which exemplified the problems of trying to convert such properties in response to the demands of caring for older people in the twenty-first century. Only a year before our visit had a stair lift been installed; it was not possible to install any other kind of lift. And only a month before our visit was the new, and first, communal dining room opened, housed in a newly built conservatory. This was not the only home to have built a dining extension; two other former private homes had also built conservatories in recent years in order to accommodate communal dining facilities for the first time.

In contrast to the small homes, which have had to extend to remain in business, some of the homes which were very large in 1958–9 have had to reduce dramatically in size. Most of the surviving former local authority homes, which in 1959 were very large, were demolished and replaced with new purpose-built homes accommodating far fewer residents in small shared or single rooms. Hillcrest, for example, an ex-PAI with 203 residents, was partially demolished in the 1970s and some of the residents were moved into a new purpose-built home on the same site which now accommodates 42 people. Another very large home in our follow-up sample adopted a rather different strategy of adapting its existing premises rather than replacing them. This was a convent, owned and run by a Roman Catholic Order. It was built in 1900 and in 1958–9 it accommodated 136 residents in large dormitories; the same building now accommodates 45 in single (mostly en suite) bedrooms.

Two of the surviving homes were purpose-built at the time of Townsend's visits. A further eight surviving homes, seven of which were former local authority homes, have subsequently been replaced by purpose-built homes. Improvement and refurbishment is of course a continuing requirement and some of these purpose-built homes, which in 1959 or more recently were very modern, are now quite the opposite. One of the homes, purpose-built in 1956, is a good example. At its official opening Pat Hornsby-Smith, representing the Ministry of Health was quoted in the local newspaper as saying that the local council had 'certainly produced, with the aid of the architect, a building which, I am sure you will agree, is a very lovely home for old people' (12 October 1956). Townsend's 1958 report on this home was similarly complementary:

> There are 7 double bedrooms, one 3-bedded room and 19 single rooms. These are extremely well furnished, in fact the best I have yet seen. ... Each resident has every amenity. ... The furniture is in light oak and the beds are low, with Dunlopillo mattresses. There are night bells in each room and lights over each bed. There are fittings for radio and electric shaver. ... In some of the rooms I saw a great deal of personal belongings.
>
> (PTC, Box 36, HI 3)

The home had no lift, however, and one had to be installed in the 1960s and then replaced in the early 2000s. Although nearly 50 years later we found that all the residents had single rooms and the kitchen and bathrooms had been upgraded, the toilet and bathing facilities were still communal and our own report on the home was not quite as positive:

> Today, its design and some of the fixtures and fittings seem outdated and the building has an institutional and rather shabby feel to it.
>
> (HR/5)

Two other homes were similarly past their best, despite being built more recently. Both were purpose-built replacements: one was opened in 1986 and the other in 1979. Both had communal bathing and toilet facilities which were badly in need of upgrading and bedrooms that were barely large enough, particularly for those requiring nursing care. One had, according to the manager, an 'old fashioned kitchen' and a laundry floor that needed replacing, and a considerable amount of money had been spent on sorting out a subsidence problem in one of its wings. Ownership of the home had recently been transferred (along with 17 other homes) from the local authority to the private sector and the new owners were planning to close it and rebuild it as a nursing home with 75 places. The other, a voluntary sector home, was discussing the

Photograph 4.2   An early post-war purpose-built home
Source: *The Last Refuge* revisited collection. © Julia Johnson

possibility of a further replacement and following our visit we learnt that it might be closed.

Another voluntary home, purpose-built in the 1930s, was clearly in a much more secure position financially and had, therefore, managed to keep abreast of required improvements. Furthermore, the quality of the original Art Deco building was superior to the new build homes of later decades. The original central block contained 12 bed-sitting rooms, offices, a drawing room, writing room and kitchen quarters. In 1946, with the help of a grant from the Nuffield Foundation, two two-storey wings each containing 12 self-contained flatlets were added. A further extension was added to one of the wings in 1970 together with a new dining room extension. Many changes had also been made in response to new access legislation: disabled equipment, widened doors, en suite bedrooms, ramps and a new lift. In the six years since 2000, there had been a painting and decorating programme, and new curtains, wallpaper and wall lights added to the communal rooms.

These data support other findings that improvement of the physical environment has been a key factor as far as survival is concerned

(Darton, 2004). We found that shared bedrooms were on the decline although not yet extinct, central heating in the 1950s was a rarity but by 2005–6 was standard, and shared washing and toilet facilities had largely (but not entirely) been replaced by en suite facilities.

Specialization and diversification

A further characteristic of the surviving homes was specialization, excluding or alternatively targeting certain categories of people. In 2005–6, four of the private homes, for example, all located in converted premises, specialized almost entirely in dementia care and one of the remaining local authority homes was in the process of varying its registration so that it could specialize in dementia care. Other homes, all purpose-built, had diversified by creating different units or wings for different categories of care. One, for example, had three units: one for nursing care, one for dementia care and one for residential care. Two others also had wings for dementia care. In contrast, there were a few homes where the managers said that that they could not cater for people with 'difficult behaviour' – another form of specialization – and there were also some homes which clearly targeted the affluent middle classes. As might be expected, such homes were located in areas where it was feasible to target this kind of market. Manor House, for example, was in a particularly affluent part of England with an above average population of retirement age.

A different kind of 'niche' market, applying in particular to many of the voluntary homes, related to specific communities of interest such as those retired from particular occupations. This could present problems, however, with some homes having to balance the need to relax and broaden their criteria for admission against the need to preserve the conditions of their charitable status. A home for ex-servicemen, for example, had in the 1980s opened its doors to ex-servicewomen, then to relatives of ex-service personnel and by 2006, when we visited, 25 per cent of residents were permitted to have no service connection at all.

Conclusion

Whether homes have survived or not is clearly dependent on a combination of national, local and personal circumstances. The fate of the former local authority homes has been very much dependent on both national policy and local politics. Townsend was disappointed in 1959 to discover that many older people were still being accommodated in buildings that had been local workhouses. Of the 39 ex-PAIs in his sample only four have survived as care homes and in all four cases the old buildings have been demolished. It is satisfying to be able to report that the era of the workhouse has ended and its legacy no longer over-shadows residential care for older people. Where property and land values are high, local authorities have been able to cash in on their capital assets, leaving the private and voluntary sectors to meet the demand for residential care. As a result of national policies, local authorities are no longer the providers of residential care but purchasers of care from the private and voluntary sectors. However, the influence of local politics remains strong. It is interesting, for example, that most of the former local authority homes in Wales are still in public ownership.

The former private homes have been affected not only by national policy and local market conditions but also by personal factors. Most of the private homes included in Townsend's sample were relatively short-lived. Many may have been on the cusp of the transformation from hotels, boarding houses or guest houses to residential care homes where the majority of the long-term residents had aged and needed care. Those homes that closed in the 1960s and 1970s were perhaps by and large temporary arrangements, providing a home and a living for their proprietors but never intended to function as lasting care home businesses. Those private homes that have survived are dependent on favourable market conditions and the enterprise and commitment of their owners. With the increasing corporatization of the private sector, however, these small businesses may struggle to survive.

The voluntary sector, as we have pointed out, has been the most stable sector. As with the private and local authority homes, homes in this sector have been affected by national policy, particularly the Care Standards Act 2000. Unlike other homes, however, they have been less subject to local markets and local politics. Rather, constraints are imposed by their charitable status and the national voluntary organiza-tions to which they belong.

Obviously for both staff and residents the closure of a home is a seri-ously disruptive if not damaging experience. In some situations, where the home is failing, closure may be the least bad option. But there are also instances where, rather than closure, what has happened is a change of ownership which has facilitated survival and continuity. Such actions depend primarily on local and personal initiative rather than national policy.

5
Residents and Staff

In this chapter we begin to compare the surviving homes as they were in 1958–9 and as they were nearly 50 years later in 2005–6. Our focus here is on who lived and worked in the homes, the residents and the staff, in the two time periods.

Our data for 2005–6 include information on the 586 older people and the 659 staff who were living and working in the 20 homes we revisited. These data were derived from the two information schedules we sent in advance to these homes, details of which were discussed in Chapter 3. Further quantitative, and qualitative, data were collected through interviews with 75 recently arrived residents and 20 home managers. The extent to which we are able to compare these data with Townsend's is limited, however, by the fact that his completed schedules relating to individual homes have not been archived. Nevertheless we are able to draw on information relating to staff and residents contained in the 1958–9 home reports and therefore to draw some conclusions about continuity and change in the population of these homes over the last 50 years.

If we compare the national demographic profiles of residential care homes in England and Wales in the late 1950s and the early 2000s, we find that there has been some considerable change in the characteristics of those living and working in them. Generalizing from his nationally representative sample, Townsend noted that 36.2 per cent of those living in residential care homes in the late 1950s were men and 63.8 per cent were women (1962, Table 16, p. 56). The men tended to be younger, however: 58 per cent of male residents compared with 46 per cent of female residents were aged under 80 years. About 40 per cent of both men and women were unmarried and well over 50 per cent widowed or divorced. Nearly 60 per cent had no surviving children, with around 20 per cent of both men and women having no known relatives (Townsend, 1962, Table 14, p. 54). Overall, Townsend demonstrated that the residents of residential care homes were 'markedly older and more infirm than the general population of old people', that they were overwhelmingly single or widowed and largely without close relatives, especially children (pp. 54–5).

By 2001, some 40 years later, the gender balance of those living in residential care homes had changed. The 2001 Census reported that 21.5 per cent of people living in residential care homes were men and 78.5 per cent women (ONS, 2005, Table S126).

To what extent are these changes mirrored in the 20 homes we revisited?

Residents: Gender, age and ethnicity

In 1958–9, six of the 20 homes were single-sex homes, one of which accommodated only men. Nearly 50 years on, one was still a single-sex home for retired professional women, and another, which had previously been for men only, still accommodated more men (60.3 per cent) than women.

Table 5.1 shows the proportions of men in each of the 11 homes for which we have full data for both time periods. It shows that the proportion of men in these homes in 1958–9 (37.5 per cent) matched the average for the time and, in line with national trends, this proportion had decreased

Table 5.1   Proportion of men by home, the follow-up study, 1958–9 and 2005–6

Home	1958–9		2005–6	
	Men (%)	No. of residents	Men (%)	No. of residents
The Brambles	0.0	43	5.8	34
Overton Court	0.0	36	0.0	51
Springfields	0.0	22	14.3	42
Mulberry House	0.0	15	26.1	23
Richmond Lodge	0.0	9	17.3	23
Acton Lodge	4.3	23	4.8	21
Newholme	32.1	28	21.0	19
Cefn Bryn	35.0	20	20.0	45
St Augustine's	36.0	25	28.6	21
St Peter's	55.1	136	31.1	45
Fairview	100.0	53	60.3	58
Total	37.5	410	22.7	382

Note: This table includes only those homes where full data on the numbers of men and women, in both 1958–9 and 2005–6 are available.

to 22.7 per cent by 2005–6. In all 20 homes in 2005–6, 26 per cent of the residents were male. By and large, however, as Table 5.1 suggests, those homes that had higher average proportions of men in 1958–9 still had higher average proportions in 2005–6. Conversely, with the exception of one small private home, those that had opened their doors to men still had substantially below average proportions of men.

It seems reasonable to conclude that in line with national trends over the last 50 years the gender balance in the surviving homes has changed, with women now outnumbering men by three to one. This can be accounted for in large part by the decline in the number of younger men being accommodated in residential care homes, particularly in the ex-PAIs where the gender balance in the late 1950s was close to 50:50. A number of the men in these institutions, as Townsend points out, had been 'prisoners or rootless seamen and wayfarers' (p. 72) who had nowhere else to live.

In 2005–6, the age of residents in the homes we visited ranged from 61 to 108 years. Table 5.2 shows the proportions of men and women in different age groups. For the follow-up homes, we do not have the equivalent data for 1958–9, so we do not know to what extent the age profile has changed over the years. Included in the table however are the data for all 173 homes in 1958–9. It could be that the surviving homes always had an older population on average than other homes in Townsend's sample. It is likely, however, although far from certain, that the proportion of men and women aged below 80 has decreased, and more so for women. It may also be the case that, in keeping with national trends, there has been a striking increase in the proportion of both men and women residents aged 90 or more. Indeed, in 2005–6,

Table 5.2   Residents in the follow-up study homes by age and sex, 1958–9* and 2005–6[†]

Age	Men		Women	
	1958–9 (%)	2005–6 (%)	1958–9 (%)	2005–6 (%)
Pensionable age – 69	12.7	6.6	9.5	0.6
70–9	45.5	20.7	36.3	11.4
80–9	39.2	48.7	45.6	45.3
90 and over	2.6	24.0	8.6	42.7
Total (= 100%)		121		465

Note: * The 1958–9 data are from Townsend 1962, Table 14, p. 54, which relate to the 173 homes he visited.
† The 2005–6 data for two of the private homes have not been included because they were incomplete. In addition, seven 'don't knows' have been deducted from the totals.

0.8 per cent of the men and 1.9 per cent of the women in the homes we visited were aged 100 or more. These data reflect national trends, in particular the closing gap in life expectancy at 65 between women and men in recent years (ONS, 2006). Despite this trend, the surviving homes in 2005–6 also accommodated a not insignificant proportion of younger men still in their sixties. This is in part due to the fact that our sample included a home for ex-servicemen and women where the median age of residents was lower than most homes in the sample.

There is little retrievable information relating to the ethnic background of residents in the archive although Townsend makes some brief references to ethnicity in the book. For example, he describes some of the residents in the ex-PAIs as being 'elderly Jamaicans, Hungarians, Poles or German Jews' (1962, p. 72). It is likely, however, that the majority of residents from black and minority ethnic groups were in ex-PAIs in the large urban conurbations. None of the four surviving ex-PAIs were in large cities, and there is no evidence in the reports for these homes that they accommodated many residents from minority groups. Of the 593 residents on which we have data for 2006 relating to ethnicity, 95 per cent were white British. Only two residents were black British and the remainder were white Irish or white from some other background, predominantly Polish. Nearly half of the residents from minority groups lived in one home in London which included a separate unit for Poles.

Residents: Family and social circumstances

Table 5.3 shows the marital status of residents in our sample. Again we do not have equivalent data for the 20 homes. Set beside Townsend's

Table 5.3 Residents in the follow-up study homes by marital status, 1958–9* and 2005–6†

Marital status	Men		Women	
	1958–9 (%)	2005–6 (%)	1958–9 (%)	2005–6 (%)
Unmarried	40.6	29.6	39.3	21.5
Married	5.3	13.0	3.0	4.7
Widowed/Divorced	54.1	57.4	57.7	73.8
Total (= 100%)		108		442

Note: * The 1958–9 data are from Townsend 1962, Table 14, p. 54, which relate to the 173 homes he visited.
† The 2005–6 data for three of the private homes have not been included because they were incomplete. In addition, one 'don't know' has been deducted from the totals.

sample, there appears to have been a slight decline in the proportions of unmarried and married men and concomitant increase in the proportion of widowed or divorced men.

Much more striking, however, and in keeping with demographic trends generally, is the decline in the proportions of unmarried women and increase in the proportion of widowed and divorced women residing in these homes in 2005–6. Within this overall picture, there are some differences between the homes in 2005–6, indicating diversity in the sample of survivors. For example, in the home for retired professional women, 47 per cent were single and in one of the former convents, five of the six men were single (retired priests). It is probable that these homes have always accommodated higher proportions of residents who never married or were single.

Table 5.4, which again compares the survivors against the whole of Townsend's sample, shows that although the proportions of men and women with no known relatives have remained much the same since 1959, the proportions with no surviving children have reduced, and for women this reduction is dramatic (from nearly two-thirds to nearly a third). This echoes the decline in the proportion of single women (see Table 5.3) resident in the homes and suggests that there has been a marked reduction in the proportion of women residents who have no close family.

In just over half the homes in 2005–6, less than a quarter of the residents did not receive monthly or more frequent visits. The three homes with the highest proportions of residents not receiving visits once a month or more were all voluntary homes which drew on very specific (and not necessarily local) communities of interest, both secular and religious. Two of them had twice the average proportion of single people residing in them and the third had more men than women.

Table 5.4 Residents in the follow-up study homes by family circumstances, 1958–9* and 2005–6†

	Men		Women		Total
	1958–9 (%)	2005–6 (%)	1958–9 (%)	2005–6 (%)	2005–6 (%)
No surviving children	57.5	45.3	60.5	31.4	34.2
No known relative	19.7	14.8	10.8	10.0	10.9
Total (= 100%)		108		442	550

Note: * The 1958–9 data are from Townsend 1962, Table 14, p. 54, which relate to the 173 homes he visited.
† The 2005–6 data for three of the private homes have not been included because they were incomplete. Data on a fourth private home have also been excluded so that Tables 5.4 and 5.5 can be compared.

In other words, these are particularly distinctive homes catering for an untypical cross-section of the older population. Many were not local people, having moved a considerable distance from former homes so potentially they were more vulnerable to isolation than the average resident. The issue of visits is an important one because Townsend argued that one of the reasons people were admitted to residential care was because of social isolation. He found that just under two-thirds of men and just over a third of women did not receive monthly or more frequent visits (1962, Table 94b, p. 518). It is possible that overall in 2005–6 residents in the 20 homes we visited were less socially isolated than the residents of these homes were in the past.

Residents: Personal assistance and capacity for self-care

Information on the needs and capacities of residents and the extent to which they depend upon assistance is inevitably subjective relying as it does on the judgements of the informant, be they the resident, a member of staff or the researcher. As anyone familiar with this line of research will know, because dependency is a relationship, measuring it is complicated (Wilkin and Thompson, 1989). To what extent, for example, is dependence on assistance induced by virtue of being in a home? We would therefore issue two cautions in reporting our findings. First, our overall data on the level of capacity of residents is crude, although comparable with Townsend's data. Second, differences we found could be as much to do with the home as actual differences between the abilities of residents in different homes. Nevertheless our findings in this area are important because a key plank of Townsend's argument was that the majority of people entering residential care were not in need of care and assistance.

Table 5.5 shows that in 2005–6, of the 596 residents on which data are available, one in five were judged to be able to go out without assistance, one in six were reported to actually go out unassisted and one in three were reported as not requiring help with dressing. Again we do not have the comparable data for the 20 homes we revisited but set against the data for the 173 homes in 1958–9, the drop in proportion of residents judged as not needing assistance is dramatic and the overall picture in 2005–6 is one of considerable frailty.

In some homes none of the residents were judged as being able to go out of the home without assistance; in others a substantial proportion were judged to be able to do so and indeed did go out without assistance. The home that reported the highest proportions needing assistance was a voluntary home: none of the 34 residents was reported as being able

Table 5.5 Residents in the follow-up study homes by need for assistance, 1958–9* and 2005–6†

	Men		Women		Total
	1958–9 (%)	2005–6 (%)	1958–9 (%)	2005–6 (%)	2005–6 (%)
Able to go out without assistance	77.3	30.6	54.2	17.5	20.1
Does not require help with dressing	82.1	33.1	72.2	28.6	29.5
Total (= 100%)		121		475	596

Note: * The 1958–9 data are from Townsend 1962, Table 94, pp. 517–8, which relate to the 173 homes he visited.
† The 2005–6 data for two private homes have not been included because they were incomplete.

to go out and only two were reported as being able to manage to dress without assistance. A further six homes similarly reported that none of the residents was able to go out, but here there were higher proportions of residents not requiring assistance with dressing. At the other extreme, another of the voluntary homes reported that just over half the residents went out without help, and a further three homes reported that more than a quarter of the residents went out unassisted. In four homes more than half did not require assistance with dressing but in four others less than a fifth could manage without help. We could not detect any factors (such as gender or tenure) to explain these differences and this suggests that the judgement of the person completing the schedule played a significant part. Nevertheless there were still not insignificant proportions of residents in 2005–6 who *were* able to manage without assistance either in going out of the home or with dressing.

As discussed in Chapter 3, Townsend developed a measure of 'incapacity for self-care' which was administered to 530 newly admitted residents who were interviewed (1962, p. 259). We likewise included his measure in our interviews with the 75 new residents in the 20 homes. Townsend's measure included 16 activities necessary for self-care which were organized into four categories: mobility and personal care, house care, communication, and activities necessary to overcome 'handicap'. Among the 530 residents interviewed at the time, Townsend found that scores ranged from 0 to 18 and he interpreted the scores as follows:

0–2   Needs little or no help to live alone in own home
3–5   Needs a slight amount of help to live alone in own home

6–8   Needs substantial help to live alone in own home
9+   Needs very substantial and continuous help from others living
      at home or close at hand.

(Townsend, 1962, p. 261)

Using this categorization, he reported that just over half the newly admitted residents scored 5 or less and he interpreted this as meaning that they could be living alone at home with comparatively little or no assistance (1962, p. 262). A quarter scored 9 or more and it was women over the age of 80 who were most likely to fall into this category (1962, Table 50, p. 263). Our data for 2005–6, as Table 5.6 shows, suggests that the profile of newly admitted residents has almost reversed: half scored 9 or more and only a quarter 5 or less. Table 5.6 also shows that men had lower scores than women and, as Townsend found, it is largely the oldest women who have the highest scores. Although the measure is crude and, as discussed in Chapter 3, difficult to administer, it is likely that we underestimated rather than overestimated the degree of incapacity experienced by the residents we interviewed because the score was based on what the resident told us and on what we felt free to ask them.

As Townsend comments, his research did not include a systematic study of 'the mental disorders of old people' (1962, pp. 271–2), though he did state that 'the question of the mental state of the elderly in institutions is one of major importance, which deserves intensive investigation' (op. cit., p. 272). He referred to differences in interpretation on the part of staff in respect of 'mental disorder' and found little evidence of training in the residential care of older people, let alone courses on mental health issues. Despite these limitations and

Table 5.6   'Incapacity for self-care' scores by sex, 1958–9[*] and 2005–6[†]

	Men		Women		Total
	1958–9 (%)	2005–6 (%)	1958–9 (%)	2005–6 (%)	2005–6 (%)
0–2	31	19	21	8	10
3–5	30	19	25	13	15
6–8	21	25	23	26	26
9 or more	18	37	31	53	49
Total (= 100%)		16		53	69

Note: [*] The 1958–9 data are from Townsend 1962, Table 50, p. 263, which relate to the 173 homes he visited.
[†] Data for six residents have not been included because they were incomplete.

in the absence of proper diagnosis, he calculated that 92 people in his sample of 530 recently arrived residents (17 per cent) were mentally impaired but only 28 (five per cent) were severely affected. Just over half of the 92 residents were restricted 'in their ability to form and communicate their thoughts to others' (op. cit., p. 268), though only 14 (15 per cent) were severely restricted.

We asked the 20 home managers to estimate the proportion of their residents who were experiencing mental health difficulties or had mental impairment. The deliberately broadly phrased question evoked a variety of responses. Three managers reported no residents in this category, but also commented that some of their residents had short-term memory loss. One of these three managers noted that active residents tended to congregate in one lounge and other residents in a separate lounge. Overall, the picture provided by nine managers was that less than 20 per cent of their residents exhibited symptoms of dementia or Alzheimer's disease, a further five managers reported between 20 and 50 per cent and the remaining six reported that over half of the residents had some kind of severe mental frailty. Five of these six described their homes as operating in segregated units with dementia wings or separate floors.

It would be wholly inappropriate to compare these figures which are based on the broad judgements of care home managers in respect of the 586 residents in 2005–6 with Townsend's estimates which were derived from his 'incapacity for self-care' measure (1962, Appendix 2, pp. 464–76) applied to the 530 recently arrived residents in 1958–9. The lack of detailed information for 1958–9 makes comparison unwise, but it is clear that one of the major discontinuities in the population of care homes is the recognition of a higher proportion of mentally frail people in 2005–6. The fact that several of the homes we visited specialized in dementia care is testimony to this.

In summary, these data suggest that in 2005–6 the surviving homes may have been serving the needs of a population with rather different characteristics from those in 1958–9. Overall, our data suggest that in 2005–6 the residents were much older and less mobile; there were fewer men, particularly younger men, fewer residents who had never married and more who were widowed or divorced. The data also suggest that far more, particularly the women, had surviving children and a greater proportion were visited at least once a month. In addition, the 'incapacity' scores for both time periods indicate that residents admitted to care homes in 2005–6 were overall far less able to manage their daily lives without substantial assistance. All this suggests that the residential care homes Townsend visited are serving a different purpose from what they were in 1958–9. However, it is important to note that, as in 1958–9, these homes in 2005–6 still accommodated a significant proportion of relatively fit older people, people Townsend would have considered not in need of residential care. This raises the same questions that Townsend raised about the purpose of residential care homes and why people enter residential care homes.

Residents: Reasons for admission

Townsend's findings led him to conclude that a substantial proportion of older people did not need to be in residential care homes, nor did they choose to be. Many were there, he argued, simply because they lacked or had lost supportive relatives or friends, because they were geographically separated from family or had had difficulties living with family and there were no alternatives by way of suitable domiciliary support services. Others had been evicted from rented accommodation and were homeless and financially insecure. He further argued that residential care did not succeed in meeting people's social needs.

We asked the 75 residents we interviewed in 2005–6 what made them decide to come to live in the home and whether they thought they had made the right decision. Just four explicitly stated that it was not their decision: 'I didn't decide. My relatives put me here' (RI/58); 'I didn't have a choice, my nephew decided for me' (RI/9); 'my sons decided. I had no voice in the matter' (RI/61); 'The powers that be said "you can't live at home"' (RI/39). At the other extreme were four residents who had made a deliberate choice to move to the home, all of them because they did not want to live alone and preferred the idea of living with others, such as one who said: 'I lived alone. I was lonely and wanted to come here' (RI/57). Between these two extremes were a great many who, with more or less resignation, said they had to come because they could no longer cope alone due to ill health or disability, or because they wanted to remain independent of their families, or because they felt lonely and insecure, sometimes following the death of their spouse.

Not all could remember the exact circumstances surrounding their arrival in the home but at least 12 had been in hospital immediately prior to their admission, ten of whom came by ambulance accompanied by no one other than the ambulance crew. Nine had transferred from another care home, either to be closer to a family member or friend, or because their former home was closing. Two had come from religious community houses and a third from a supported group home for people with mental health difficulties. Information we were able to gain about how much help and support was received by the 61 who had come from their own home directly or via hospital was patchy. Just over two-thirds had been owner occupiers. Nineteen said they had not received any help from people other than family, some of whom may simply have been reluctant to admit that they had had help or that they could not remember, as a further 12 confessed. The remaining 30 described a variety of arrangements ranging from having a 'cleaner' or a gardener to home care three times a day, seven days a week, and including meal delivery and help with bathing or showering and other kinds of personal care.

Regarding the question of whether the right decision had been made, just two were unambiguously unhappy about the decision and expressed dissatisfaction with the home. Both had been transferred from another home and neither had been consulted about the move. The majority, however, expressed more ambivalence saying they 'thought' they had made the right decision, that they liked it but hoped they would be 'going home soon' or that the home was simply 'alright'. Others acknowledged that they had had little option: 'it was the only thing to do really' (RI/68); 'I'm happy. My son has no worries now' (RI/32). Some were relieved to have left behind them a home of their own which had begun to feel unsafe.

> The reason I left home: I had no fencing, open gardens, I had to walk down a wide path – so if I was tottery, there was nothing to hang onto. So I felt nervous. I was getting nervous about walking and shopping and being at night on my own. There were people knocking on doors at night – even though my flat was upstairs. I didn't open the door ... people saying they were from the gas company.
>
> (RI/35)

Expressing what Peace et al. describe as 'a distinct trade-off between decreased autonomy and more security' (2006, p.123), the wife of a married couple who were still getting used to their transition to a care home commented:

> We would still prefer to live in our bungalow, but we feel safe in here. If anything were to happen, there's somebody close at hand. It feels nice. At our age, I'm glad we're with somebody. It makes you feel better. As long as you hear voices, it gives you comfort.
>
> (RI/51)

There was still a substantial minority (24 of the 75 residents interviewed), however, who were unambiguously positive about the decision that had been made saying it was 'definitely' the right thing to do, often adding a comment such as 'the nicest home I've ever seen' (RI/59); 'I can't believe my luck' (RI/20) 'I enjoy it here. I love it' (RI/28).

In regard to whether the homes we visited in 2005–6 were able to meet the social needs of residents we, like Townsend, asked the 75 people we interviewed whether they felt lonely in the home (often, sometimes or never). Townsend found that 46.7 per cent of those interviewed said that they were often or sometimes lonely (Townsend, 1962, p. 350) and we found a not dissimilar proportion, 43.5 per cent offering the same answers. Townsend argued that this proportion was much higher than found in samples of older people living in the community, for example his own community study of Bethnal Green (Townsend, 1957) or Sheldon's study of Wolverhampton (1948). Although this loneliness, he observed, often arose as a result of loss of a spouse or other close relative, 'separation from relatives and growing infirmity, as well as the shortcomings of institutional life, often reinforced such feelings' (1962, p. 351). Our findings were broadly similar although only a few of those who said they were often or sometimes lonely specifically mentioned a recent bereavement. One man, for example, said 'Sometimes I go to bed at night and just cry. My wife Evelyn died only 19 months ago. We were married for 68 years and it turned my life upside down' (RI/69) and another said 'Yes. I miss my wife very much. ... I miss her terribly' (RI/8). Our findings suggest, however, that 'separation from family, growing infirmity and the shortcomings of institutional life' identified by Townsend might be important factors. Indeed, although a more recent community-based study of loneliness (Victor et al., 2005), using a similar single-question self-rating, found a higher proportion of older people rating themselves as often or sometimes lonely than in the earlier community studies of Sheldon and Townsend, this proportion was still less than we found in the care homes we visited.[1]

The kinds of factors we came across, which seemed to be associated with those reporting loneliness, included a longing to return home, missing family or friends, difficulties in communicating with, or finding something in common with, other residents or social isolation due to impairments, including sensory impairments. As Townsend recognized, social isolation and loneliness are not the same thing and some could be lonely in spite of caring friends or staff. For example, a resident aged 101, when asked if he was lonely, said:

A little bit. Sometimes in the middle of the night. When I take my clothes off and get into bed. Yes, just for a second. Then I think

I can't be lonely – they call me in the night 'Are you alright Fred?' They call me to see if I'm alright. And there's one woman here, Gwenllian's her name, and she always calls 'goodnight Fred' and I call 'goodnight'. And it's nice. It doesn't work wonders, but it's nice to be recognized.

(RI/44)

Conversely, of course, there were those who were socially isolated, or social isolates, but not lonely, as one woman put it, 'I'm alone, but not lonely. I'm used to being alone' (RI/33). Townsend noted that those who had been used to living alone were less likely to feel lonely. Of the 34 who told us in 2005–6 that they were never lonely, 11 had never married, 17 were widowed and a further four divorced, nearly all for ten years or more and some for as many as 30 or 40 years, and two were a married couple living in the same home. Almost all these people were strident in their responses: 'I can honestly say I have never felt lonely' (RI/38); 'I like being alone' (RI/31); 'I'm not lonely, not bored, perfectly happy' (RI/26); 'I've too many things to think about' (RI/23); 'No ... I've always liked my own company' (RI/6); 'No. I'm not a lonely sort of person. It doesn't bother me very much' (RI/68); 'Oh goodness no! There's no time for that' (RI/67); 'Never. It's your own fault if you are' (RI/65). Just under half of those who said they were never lonely had low incapacity for self-care scores, indicating that they needed little or a slight amount of help. The rest, however, needed substantial amounts of help, ten of whom required very substantial help. Increasing infirmity therefore is not necessarily associated with feeling lonely. Townsend gave little attention to the majority who said they were never lonely, but his data and ours suggest that, as well as being associated with marital status, there is perhaps a personality factor here. Quite simply, some cope with the shortcomings of institutional life and the challenges of old age more easily than others.

It is important to point out too, that there were a few residents both in the 1950s and the early 2000s for whom moving into residential care had relieved their loneliness. For example, one of our interviewees, who claimed never to be lonely, said that although she was not lonely in the home she had been lonely in her former flat (RI/49).

Staff: Numbers and roles

It is difficult to generalize about staffing in 1958–9, either from Townsend's book or from the archived reports, because neither contains any systematic data on this topic. However, there is sufficient material in the archive to amplify on the very varied range and type of staff in different kinds of homes in the late 1950s.

For example, an ex-PAI which accommodated 203 residents, employed a superintendent, assistant superintendent, matron and assistant matron (both State Registered Nurses (SRN)), all of whom were resident. It also employed four stokers, two engineers, nine laundry staff, six kitchen staff, 16 domestics, two seamstresses, a carpenter, a painter, two gardeners, two clerical staff, a part-time organist, two chaplains and a chiropodist. In addition, it employed 29 attendants, six of whom were men, three were State Enrolled Auxiliary Nurses (SEANs) and all were aged between 20 and 60.

By way of contrast, a small private home accommodating 15 women was owned and run by matron and her husband (who did all the odd jobs). They lived in the home and employed three full-time resident female assistants who were described by matron as 'fallen girls from a home for high grade mental defectives', a part-time washer-up and a full-time assistant matron. Not all the private homes were so adequately staffed, however. Another for ten residents, had four resident staff: the proprietor and his wife, his 70-year-old aunt and a 33-year-old male nurse described as doing 'about a third of the work of a full time person, and spends the rest of his time helping in a relative's printing business. He is said to be dying of cancer.' Apart from these staff, there were two cleaners and an SRN who called weekly 'to bath any who can afford to pay for it' and sometimes did some night nursing. These kinds of arrangements were not untypical of the private homes in 1958–9.

The staffing of the voluntary homes provided further contrasts and diversity. One of the surviving Catholic convents, which accommodated 23 older residents in 1958–9, six of whom were retired priests, was staffed by 11 resident sisters who did everything including the redecorating. In some of the voluntary homes, the shortage of paid staff meant that residents were often engaged in domestic activities which were central to the running of the home such as laying tables, making beds, growing and preparing vegetables – an issue we return to in Chapter 7. Overall, however, the voluntary homes had higher staff/resident ratios than the local authority homes, although many were very poorly paid and there was a shortage of qualified nurses (Townsend, 1962, pp. 160–2). Fairview, a secular voluntary home for 53 men, employed 28 staff. Ten were resident including the matron, the secretary, five untrained 'nursing orderlies' and three 'maids'. The 18 non-resident staff included four trained nurses, five domestics, two porter cleaners, a driver, three gardeners, a boiler-handyman, a chef and a kitchen porter.

Photograph 5.1 'Matron' and staff 1959 and 2005
Source: The Peter Townsend collection; *The Last Refuge* revisited collection. © Jean Corston (top), © Julia Johnson (bottom)

It is probably true to say that although diversity in staffing arrangements among the surviving homes still existed in 2005–6, the way in which homes were staffed was much more standardized, not least because of the requirements regarding resident staff ratios, training and qualifications of the then CSCI in England and the then CSIW.

We collected information on 659 members of staff who worked in 18 of the 20 homes. Many of them worked part-time hours and some homes, at the time of our visits, were employing agency staff and/or bank staff as well. Reflecting the changed function of the care homes, the categories of staff employed across them had to some extent changed since the late 1950s. The wardens and matrons have gone and have been replaced by managers, although it was far from uncommon to hear the women managers in 2005–6 being referred to as 'matron'. In addition to the managerial staff, the 2005–6 staff included administrative and clerical staff, qualified nurses, senior care staff, care staff, domestics and kitchen staff, drivers and handypersons.

Exemplifying these changes, Fairview (mentioned above) employed three times as many staff as in the past, although it catered for only five more residents. Of its 82 staff in 2006, 83 per cent were women, all 'white/British' with the exception of one 'other/white'. Table 5.7 shows the breakdown of its staff by category, each distinguished by a uniform: the nurses wore hospital blues, the care staff polo shirts and navy trousers, the domestics blue and white-striped tunics, the deputy matron a green dress, and the manager and office staff no uniform at all. Ten of the staff were 'bank' staff, employed as and when needed, and the remainder were on permanent full-time or part-time contracts. The normal full-time contract was 35 hours, but five members of staff had long-standing contracts for a 40-hour week. Part-time contracts ranged from ten to 30 hours a week. The day shifts for care staff were from 7 a.m. to 2 p.m. and from 2 p.m. to 9 p.m. At night there were four care staff and a nurse on duty. The overall median age of the staff was 49 and only 14 were aged under 30, all bar one being care staff. The care staff were

Table 5.7   Number of staff at Fairview, by category and sex, 2006

	Male	Female	Total
Nursing staff	2	11	13
Care staff	8	36	44
Domestic, cooks and maintenance staff	5	17	22
Office staff	0	3	3
Total	15	67	82

significantly younger: their median age being 42 compared with 56 for the nurses and 57 for the domestic staff. Nine of the domestic staff were aged 60 or over and the oldest member of staff, a carer, was 69.

These characteristics were typical of the 20 homes in general: 86 per cent of the 659 staff were women and nearly three-quarters were aged between 30 and 60. A very small percentage were under 18, mainly school pupils doing Saturday work in the kitchens. A further small percentage were over 60 with one or two over 70. There was greater ethnic and racial diversity among staff than residents. Although the great majority (86 per cent) were white British or Irish, 4.1 per cent were from an 'other' white background and 6.4 per cent were black or black British (African or Caribbean) or other black background. The 2.9 per cent Asian or Asian British were predominantly from an Indian or Chinese background and none were of Pakistani or Bangladeshi origin. This profile seems to be consistent with the national profile for the social care workforce in care homes, in England at least, in 2006–7.[2]

The majority of workers from minority ethnic groups were concentrated in just a few homes in London and the Home Counties. For example, a private home for 37 residents, employed 23 members of staff (17 women and six men) in 2006. The manager, who was Indian, employed a diverse group of carers: five were black; four Asian, including Indian and Thai; eight non-British white, mainly Polish; and four white British. There was one Chinese woman whose aim was to improve her English, pass her adaptation exam for English nurse training and then join a big hospital. In fact, she was the only member of staff who was a qualified nurse; four more had National Vocational Qualifications (NVQs), and the others had undertaken some basic training as carers. The first company to take over ownership of this home, which had been a family business in the late 1950s, had a policy of recruiting Indian nurses, who formed 95 per cent of its staff, and then Filipino nurses, who hoped to move on to hospitals and nursing homes after passing their adaptation exams.[3]

The managers of 13 of the 20 homes we revisited in 2005–6 reported turnover of care staff as low and 14 managers reported low turnover of domestic and kitchen staff. Although none reported high turnover of care staff, seven managers said they faced some difficulties in care staff turnover and four had some problems with domestic/kitchen staff turnover. Low pay was the most frequently mentioned concern and this tended to be linked to the recruitment and retention of staff.[4] Sickness and holiday entitlements added to the pressures on managers, as did externally required changes in the demands on staff, such as training requirements. Three managers reported that transport problems affected recruitment. There were references to personality clashes in the staff group in one home and to punctuality issues in another. The manager of one faith-based home summed up the difficulties as follows:

> The main problem is recruiting staff. It is a low paid job and not highly thought of. Promotion prospects are poor. Quite a few of the staff recruited are young, who then move on into nursing. Retention is a big issue. There is a high rate of absence due to sickness mainly related to stress.
>
> (MI/6)

However, it is important to re-emphasize that these problems should be seen in the context of no reported high turnover of care staff and very little high turnover of domestic/kitchen staff. Even when pay was recognized to be low, it was not always the case that this was accompanied by recruitment problems. A local authority home and a voluntary home, for example, reported that pay was better than for those working in the private sector in the same area, and in a further two cases, where the homes were run by Orders of nuns, the issue of pay related only to the secular care and domestic staff. It is also worth noting that in certain circumstances the distinction between care and domestic staff can be blurred, not least in respect of night staff. In describing the latter's duties, half of the managers mentioned housekeeping work. In one privately run home of 26 residents there was 'some cleaning of toilets, some kitchen deep cleaning, some vegetable preparation, the bulk of the laundry' (MI/14). Agency staff, as opposed to bank staff, were in general not regarded as of consistently good quality, though they were at times used to ease staffing crises, despite being expensive, which in turn caused resentment among the permanent staff.

So how does this staffing profile for 2005–6 compare with the situation in the late 1950s when Townsend and his colleagues were doing their fieldwork? The contrast is great, if not surprising, though it is not possible to discern a consistent pattern across all types of homes. For instance, although the majority of homes have increased their staffing complement, even when the number of residents they now cater for has decreased, this was not always the case. The ex-PAI for 203 residents replaced by a 1970s purpose-built home for 42, had a somewhat reduced staff complement in the early 2000s compared with the late 1950s. This is most likely, however, a reflection of the dramatic reduction in capacity. For the most part, capacity had increased and alongside it staff numbers. Furthermore, the increase in staff has consistently outstripped the increase in residents. So, for example, while two of the private homes had nearly double the number of places when compared with the 1950s, they had tripled in staff numbers. In another private home, although the number of places had not increased, the staff complement had more than doubled. One of the local authority homes, purpose-built shortly before Townsend visited, had through internal conversions reduced the number of places from 36 to 32, but the total number of staff had increased from eight to 30. Unfortunately, the data available on the homes in the 1950s do not enable us to make full-time equivalent comparisons in the two time periods and therefore to judge the extent to which the staff/resident ratio has changed over time. There is no doubt, however, certainly in the private homes and some of the voluntary homes, that many of the staff in the 1950s worked much longer hours than they did in 2005–6, simply because they were resident in the homes.

Sixteen of the 20 homes were explicitly referred to as having residential staff in 1958–9 and this total went down to five in 2005–6. Three of these five were small private sector homes and the other two were voluntary sector homes run by Orders of nuns. In respect of the latter, whereas the nuns had done most of the work in the 1950s, there was in 2005–6 a large contingent of lay staff. One Mother Superior said, 'It was a matter for regret that two years previously it had become necessary to employ a qualified chef, and, for the first time since 1920, there was no Sister in the kitchen' (MI/12). In the other home run by nuns, catering had been handed over to an outside company and all services like laundry, sewing and reception had been taken over by lay staff and much of the caring was being done by 40 care assistants. One of the reasons for this changed pattern of staffing was the ageing of many of the nuns and the shortage of younger nuns coming in. In addition, however, there has been the introduction of standards, including training to obtain qualifications.

Staff: Training and qualifications

Townsend's broad conclusion was that there was in the late 1950s a woeful lack of training instruction in caring for elderly and frail people and he recommended that there should be at least one qualified full-time nurse for every four residents and that the person in charge of the staff should have a social science diploma or should have completed a course in social medicine and welfare (Townsend, 1962, p. 421). For over half of the homes visited, there was no record in the available documentation from 1958–9 of qualifications or training. Two matrons were reported as having run a mother and baby home in the past, and another had followed a course run by the national Old People's Welfare Committee. Staff with nursing qualifications was reported to be employed on either a full-time or part-time basis in the remaining six homes. The contrast with 2005–6 is very marked.

In all but four of the 20 homes visited in 2005–6, reference was made to all or the majority of staff trained to NVQ Level 2 or beyond. Two small privately run homes did not provide clear details, though the manager of one of them said that the cost of training was a problem: £1500 had been spent in two months. The manager of a faith-based home said that the NVQ requirements put some people off, and that those who were trained were liable to move on to a nursing career. In another faith-based home, the older nuns had found adapting to requirements very difficult. Another manager from the voluntary sector had reached agreement with the CSCI that long-serving care assistants would not have to undergo the training unless they wanted to, but all new recruits would have to gain a recognized qualification.

Regarding the content of training, frequent reference was made by the home managers in 2005–6 to internally run courses covering a huge variety of topics (many overlapping with NVQ requirements). They included health and safety, manual handling, food hygiene, fire precautions, infection control, dementia awareness, mental health, incontinence training, medication, death and dying, bereavement, cremation and burial and palliative care. When asked specifically whether staff were trained to work with people with mental health difficulties, five managers said that training was organized in the home. A further five sent staff to local colleges and, in one home, both internal and external courses were available. The focus was in the main on dementia awareness and dementia care – one home used the Bradford Dementia Care Mapping training package – but managers also referred to courses on memory loss, and on bizarre and challenging behaviour. One manager commented that 'staff have learnt that if there is aggression, it means people don't like what they're doing' (MI/8). Another remarked 'staff find it difficult to deal with challenging behaviour. What is the problem? It needs time to address it. We don't have the time' (MI/19).

Townsend's emphasis on nursing qualifications for care staff has not been followed through, though at least nine of the homes employed people with nursing qualifications, in some cases highly qualified nurses from overseas. In fact, the managers of four homes, all located in London and the Home Counties, said they were dependent (at least in part) on people from outside the UK to staff their homes adequately. Three of them were private sector homes that ploughed profits back for further development. One of these, as mentioned above, had a long tradition of employing workers from overseas: as we described in Chapter 4, before it was sold to a public company, it belonged to a retired medical missionary and offered accommodation to 'foreign girls' who helped out in the home (PTC, Box 36, HI 7).

Managerial responsibilities

Despite the increasing emphasis on management and administration, few managers had received training in this specific area of work or had been recruited because they had the relevant skills. Comparing Townsend's reports with our own interviews with managers, it is clear that in 2005–6 there were much greater and very different expectations placed on managerial staff than in the late 1950s, mainly as a result of legislation in the intervening years which demands far stronger accountability. Interestingly, in one of the local authority homes among those we visited, the manager was relieved of a substantial amount of financial administration by personnel at County Hall. In addition, she was able to use the independent inspection process to obtain resources for the home from the local authority. In contrast, some of the small private businesses had to shoulder the entire burden of management and administration ranging from hiring and firing staff to building new extensions and ordering supplies for the home. One manager of a small private home had been facing staffing difficulties and, as a result, had been doing the cooking herself. She was unable to find time to provide us with statistical information on the staff and residents. The manager of one of the voluntary homes summed up the competing demands on her time as

> being on top of health and safety legislation and employment law; being a nurse; being the figure head. One minute you're talking to a resident and the next to the handyman about radiator valves.
>
> (MI/13)

Adding to this, the researcher noted,

> [t]he second day I visited, she was being visited by her line manager from headquarters who was already there when I arrived at 9.30 am and was still there when I left at 5.10 pm.
>
> (HR/13)

Photograph 5.2   Manager's office in 2006
Source: *The Last Refuge* revisited collection. © Sheena Rolph

The demands placed on managers by the owners was well illustrated in the case of the private home which up until the 1980s had been run as a small family concern. Since then it has been sold several times, until it was bought in 2005 by one of the largest profit-making care home operators in the UK. Two wings had been added to the original home in the 1980s and the staffing had more than doubled between 1958–9 and 2005–6. The immediate impact of the recent 'corporatization' was an increase in formal procedures, which brought

> dramatic changes to the way the home functions and runs ... there have been large financial changes ... if the manager wanted anything for the home when it was run by (a previous company), she just used to ring through a request. This would usually be approved if the case was made, and the company would send the items quickly. Now ... there is a capital expenditure form. The budget is agreed and set up in the previous year, so there must be forward planning. There is now much more paperwork and there are many more people to contact during the process.
>
> (HR/17)

This manager was far from alone in referring to increased formality and less of a family feeling. Time spent on administration meant less time spent with residents (and staff). Some managers regretted this. While the notion of 'being part of a family' was common in the homes visited in 1958–9, it was clear that the task of caring for substantial numbers of frail, older people in 2005–6 had resulted in more of a business culture. This is not to deny the genuine commitment of many staff, but the pressure of time and a tight staffing complement as well as the need to cater for a relatively large number of people to make ends meet have led to a focus on the efficient running of a home rather than on the quality of life of the residents.

Conclusion

As we have indicated, the 1958–9 data available to us on the residents and staff of the 20 homes we revisited in 2005–6 is to an extent limited and therefore we must be cautious about the conclusions we can draw about continuity and change in regard to who lived and worked in the homes in the two time periods. That said, our analysis suggests a number of significant changes in the populations of the homes.

Overall, in 2005–6 there are a greater proportion of women residents in homes we revisited and the evidence we have suggests that the residents are older and more incapacitated than they were in the past. While several homes, particularly the voluntary homes, continue to cater for certain secular and religious groups of people, overall we have concluded that these homes are performing a rather different function when compared with that of 50 years ago. Residents require a greater amount of personal care and support although there are still those who are relatively independent. Nevertheless it remains the case that people come to live in a care home not so much because they choose to but because they recognize that their options are very limited.

To meet their changing function, staff numbers have increased substantially over the years although many, in 2005–6, worked part-time hours. The workforce, particularly in the private and voluntary homes, was much more highly differentiated than in the past. It was also more closely regulated, although there was still scope for exploiting cheap labour. Like the residents, the great majority of staff were women, making these care homes more gendered contexts than they were in the past. As such there is a problem of care home work being undervalued, so although the staffing profile in 2005–6 was relatively stable, there were problems of pay, recruitment, retention, sickness and the quality of agency staff. For some, working in the homes was simply a means to an end. Training was much higher on the agenda than in the late 1950s, but it seemed to focus on practical issues such as first aid and food hygiene rather than on everyday social care, though there were opportunities to address mental health issues.

Paradoxically, given Townsend's concerns, the increase in staff numbers and the formalization of procedures has led to many of the homes having a more institutional feel than in the past.

6
The Living Environment

In this chapter we address the question of what kind of living environ-ments the 20 homes provided in the early 2000s and how they compare with the past. Forty years ago, Pincus adopted the term the 'institutional environment' which he defined as the

> psycho-social milieu in which the residents live as expressed through and/or generated by (a) physical aspects of the setting, including design, location, furnishing and equipment; (b) rules, regulations and program which govern daily life; and (c) staff behavior with residents.
>
> (1968, p. 207)

Drawing on our home reports and those written by Townsend and coll-eagues in the late 1950s, we examine the relationship between these aspects of the home environment and the lives of residents, and the extent to which this has changed over time. We begin by looking at changes to the physical environment and then compare the psychosocial environ-ments in terms of rules and routines and staff attitudes and practices.

In Chapter 4, we described some of the considerable changes since 1958–9 that have taken place in the buildings themselves, both interiors and exteriors, through extensions, additions and sometimes wholescale refurbishment. Ex-PAIs have been closed down, but some homes have been rebuilt on site or nearby, pointing up the contrast between what was acceptable in the two periods. We found that the architecture of the homes in 2005–6, their siting in relation to the local community and their surroundings, all had an impact on the daily life of the residents, whether the homes were purpose-built or converted houses dating sometimes from the Victorian or Edwardian periods.

Location and scale

Townsend thought that the size of the home, together with its distance from the home localities of the residents, sometimes had a direct impact on the feelings of isolation experienced by the residents. For example, it was observed that many of the older men in Hillcrest, one of the ex-PAIs we revisited,

> seemed to feel cruelly uprooted and quite unattached to their present surroundings. One old man said, near to tears, 'I was thinking only this morning of my village ... and it seems to draw me; it draws me all the time – 'tis only natural, isn't it?' He put out an arm and made a tense pulling movement to illustrate this feeling. Another described how he sits at the roundabout watching the coaches coming up from the West Country and how sometimes he sees one which comes from his old home, but by the time he has read the name of the proprietor it has gone and he hasn't had time to search for a familiar face.
>
> (PTC, Box 36, HI 8)

Like Hillcrest, several of the surviving homes, particularly the very large ones, had in the 1950s a wide catchment area meaning that many residents had been uprooted and separated from familiar places and people. In addition, however, Townsend was critical of the remoteness of some of the homes from local communities and facilities. Large grounds, he argued, could be 'disconcerting if not somewhat frightening' (Townsend, 1962, p. 110). One matron was quoted as saying that '[t]he council always seems to buy a Home right out in the wilds and when it gets dark you feel cut off from the world. You feel right off the map' (p. 111). None of the surviving homes which we revisited in 2005–6 were located 'out in the wilds' although a few had been more remote in the past than they are today. The most extreme example was a voluntary home which in the late 1950s was about one-and-a-half miles from the nearest village and eight miles from the nearest town. It was situated in 62 acres of parkland with a drive nearly a mile long. The isolation engendered by the large grounds and the remoteness from the neighbouring town meant that this was almost an enclosed home, with few visitors or outings, and in this way the size of the grounds had a negative impact on daily life. As mentioned in Chapter 3, by the mid-1980s it was recognized that the location of this home (as well as the building itself) was no longer appropriate as a residential care facility. Consequently, a new home was built to which all the residents and most of the staff were moved. The new home, which one of us visited in 2006, was located in a residential road close to the centre of a small market town some 40 miles from the original home. Although the residents at the time were once again uprooted, the new home was within easy reach of local shops and the wider community. Indeed when a new manager was appointed after the move, the tall leylandii bushes in front of the home were cut down to make it more visible and a part of the local community.

Apart from the reduction in the remoteness of some homes, there have been substantial changes to the scale and size of many of them and the facilities they offer. The 'appalling inadequacy' of Hillcrest in the late 1950s was reported to contribute to the 'utter desolation' many of the men felt as revealed in their interviews. Some of the 'fairly grim' rooms were found in this ex-PAI in 1959:

> On the ground floor of the Main Block is a ward for 8 men. It contained nothing but iron beds, 8 metal lockers and 9 bedside mats. There were no signs of personal possessions about. One would not have thought that anyone lived in this room. It was completely bare. [The Superintendent] expressed dissatisfaction about the 'deplorable' over-crowding here. ... The space between these beds was about two and a half feet.
>
> (PTC, Box 36, HI 8)

This building, described with such feeling by Townsend in 1959, no longer exists, and in 2006 the replacement home presented a very different scene:

> [A]n attractive 1970s two-storey building with gable ends, which is built around a semi-enclosed garden. The entrance to the home is through French doors into a bright foyer. ... There are five small lounges ... and two dining rooms ... and 17 single bedrooms. Residents can bring their own furniture if they wish and many rooms were full of personal possessions ... one room, larger than most, contained a double bed, a mirror in a gilt frame, attractive bedside lamp, radio and clock and a pot plant. The cupboard doors were covered with family photographs.
>
> (HR/1)

Although in this case the PAI was replaced by a more attractive smaller building, more domestic in style and size, as we mentioned in Chapter 4, other homes have expanded since the 1950s. Paradoxically, therefore, some homes which were small family-sized homes in the late 1950s have grown to institutional dimensions over time. This is particularly the case for the former private homes, several of which have tripled or quadrupled in size. Many lacked communal facilities in 1958–9, such as a dining room or a lounge, which had consequences for how the daily life of residents was conducted. For example, in 1959, one of the former private homes that has doubled in size, had no lounge or dayroom and there was nowhere for residents to congregate. Consequently, as one resident then reported, 'You've got to be very careful. You're kept to your own rooms' (PTC, Box 36, HI 10). The opposite was true in 2006. The home now has two very large lounges and a policy of more or less requiring those residents who are able to come down to use them during the day so as to conform to a belief in the home as a community. In 1959 the bedrooms in this home were described as being 'all fully equipped as bedsitters' (ibid.). Today's bedrooms are similarly furnished – with personal effects on display as in 1959 – but many in the new wing are quite small and cramped and it would be difficult to describe them as bedsitters. Whereas in 1959, therefore, the bedsitting rooms were the focus of daily life for residents, by 2006, the 'heart of the home' had shifted – in the manager's mind at least – to the large lounge.

Internal spaces

Segregation

Both now and in the past, the built environment has played a role in the segregation and categorization of residents. In 1958–9, Townsend found that 'the lay-out and amenities of the buildings and the division of duties among the staff depended on the principle of segregation' (Townsend, 1962, p. 103). In the ex-PAIs, residents were often segregated according to gender, and also according to whether they were 'healthy', 'infirm', or 'chronic sick' (p. 73). In Hillcrest, for example, men and women slept in separate blocks but shared the same dining room and some of the day rooms. Gender segregation, where it existed in 1958–9 in the other types of local authority, private and voluntary homes, was generally achieved by accommodating men and women in different (single-sex) homes. However, occasionally residents were separated within the same home. In one of the convents, for example, female and lay residents were separated from the priests' quarters by means of 'curtains across the corridors' (PTC, Box 36, HI 2). Segregation affected the style, atmosphere and daily activities of the homes, and several matrons in 1958–9 confided to the researchers that a mixed home would be preferable.

Although, in 2005–6, one single-sex home remained, segregation of the sexes in the mixed gender homes had more or less disappeared. Segregation according to functional ability, however, continued. Sometimes this was planned for, as in the later purpose-built homes, such as Cefn Bryn which had two dementia care wings on the ground floor and two residential care wings on the upper floor, and Fairview which had three separate self-contained wings for residential care, nursing care and dementia care, respectively, the latter being secured by keypad accessible doors. But sometimes the use of space for segregation was more subtle. For example, there was a tradition in one of the homes whereby the more able residents 'were directed' upstairs so that they could spend their time on the self-contained top floor of the home This included a sitting room, small dining room and a kitchenette. According to the manager, those who used this facility were 'self-selecting' and only came downstairs for quizzes (MI/8). Clearly, it was a strategy used to separate themselves from those with cognitive or other impairments.

Corridors and thresholds

A further contrast in the use of internal spaces between the two periods related to corridors which over the years have been transformed from cold, utilitarian passages, as depicted in some of Townsend's photographs, to social spaces. The introduction of central heating is a key reason for corridors becoming more hospitable. In addition, by the early 2000s considerable importance was being attached to way-finding, particularly in design for people with dementia, where the emphasis is upon the need to enhance corridors and circulation areas by means of good lighting, a lively choice of colour schemes and technical aids for those with impairments (Passini, 1992; Duffin, 2008).

Several of the photographs in *The Last Refuge* feature the long, bare corridors with hard floors and undecorated stone walls which often characterized the ex-PAIs. Like the inner streets of a hospital, they were used to carry laundry, transport goods and medical supplies. Three of Townsend's unpublished but archived photographs depict corridors stretching to a central vanishing point many metres away with tiny figures at the far end indicating the vastness of scale. They tell the story of daily lives lived in a hostile built environment, which is difficult for older or less able people to negotiate. In the 1958 home report on one

Photograph 6.1   Two entrance halls 2005–6
Source: *The Last Refuge* revisited collection. © Sheena Rolph (top and bottom)

voluntary home, for example, it was said that '[t]he passages were rather long and cold and some of the residents have to go as much as 20 yards to the lavatory' (PTC, Box 36, HI 1); in another 'there was a long way to walk to the bathroom, thirty to thirty-five feet' (PTC, Box 36, HI 2). Even in the newly constructed homes visited by Townsend, purpose-built after the war, there was often a repetition of 'lengthy corridors' which 'tended to recreate the atmosphere and severity of an institution' (1962, p. 113). Among the surviving homes, there is just one report entry from 1958–9 which casts corridors in a more favourable light:

> It was pleasant in a convent to hear some of the younger Sisters shouting and calling to each other along the corridor. Often in the convent homes there is a prevailing silence which seems to be reluctantly broken when anyone speaks – a sort of paralysing air of sanctity.
>
> (PTC, Box 36, HI 2)

In 2005–6 we found that in all the purpose-built homes, built to replace the institutions Townsend had visited, the lengthy corridors were still being repeated, creating an institutional feel. Nevertheless we also found that corridors and other circulation spaces, in both the purpose-built and converted homes, sometimes had more positive social uses. Many had been made more congenial and attractive through the use of heating, carpets and decoration. As a result, we found that they could be important sites for daily social interactions for some residents, offering opportunities for exercise and chance meetings. Peace et al. (2006) found that moving out from 'anchor points' is important for people in their own homes, and that moving between points – from one room to the next, or into the garden – has physical and psychological benefits, providing variety in the daily routine, and also a sense of identity and independence. Our findings suggest that this holds true for many care home residents who made similar efforts to maintain autonomy and identity by establishing safe and beneficial daily routes and border crossings to and from the anchor point of their own room. Although single rooms now offer much-needed privacy, they can also be lonely and quiet. Venturing beyond the room, crossing the border into the corridors can open up the world of the home: noise, work, trolleys and welcoming greetings from, or short conversations with, other residents or staff. As one resident noted in her diary, she '[w]ent for a walk along our passages' and '[m]et a carer collecting residential's supper trays' (RD/13). Another, although limited in her movements, went 'up and

Photograph 6.2 Sitting on the threshold 1959 (above) and 2006 (below)
Source: The Peter Townsend Collection; *The Last Refuge* revisited collection. © Jean Corston (top), © Sheena Rolph (bottom)

down the corridor' for her bad back, as well as for social reasons (RI/1). Likewise, corridor walking was part of the daily routine for the resident who wrote:

> Dressed and ready by 9am. Eat apple. Read until 10. Woman's Hour from 10–11. Walk up and down corridor – then read. Lunch arrives at 12.15–12.30.
>
> (RD/19)

Corridors and circulation areas are also important for those diagnosed with dementia who need to walk – sometimes constantly and repetitively – around the home (Marshall, 2004). In one home we noted that 'the dementia care residents spent their waking hours in the two downstairs lounges or walking the wide and spacious corridors downstairs' (HR/6). However, corridors could still, as we observed, pose problems for some residents, wandering, lost and uncertain as to where to go. One interviewee described his role in guiding those who were lost back to the lounge or to the central part of the home.

We also found examples of residents occupying corridor space for their own personal use. One man, an organist, regularly sat in an easy chair in the corridor studying the sheet music for his next organ piece. He greeted, and was greeted by, passers-by and seemed to enjoy the sociable atmosphere. Another had personalized an area of the corridor with her own table, flowers and photographs.

In several homes there was a quite particular use of thresholds by residents who, as in 1958–9, tended to congregate just inside the glass front doors, looking at the outside world across a kind of borderland, symbolic perhaps of the broader segregation of older people from society. In one 'traditional' Welsh home, the men who never sat with the women in any of its three lounges, sat in the hall with a view of the front door. As the assistant manager pointed out, while the women might join them, it would never be the other way round. Thresholds, as well as larger lounges, could form the heart of a home – and in some homes there was a regular and obviously affectionate use of small spaces colonized and adapted to their own use by one resident or groups of residents who had formed friendships.

> When I visited, two women were sitting companionably on the wide landing outside the dining room and were deep in conversation, looking out over the balcony with its wrought iron chairs and table; another was sitting just inside the dining room in an easy chair,

next to a table with a huge vase of flowers (sent by her family), a music centre and her wedding photo. The atmosphere was that of a self contained living area or flat, where friends lived together, and moved out of their rooms to colonise more of the home with their own flowers, pot plants and photos.

(HR/8)

These examples of how the public spaces in the homes were used in 2005–6 suggest that, over time, residents find ways of dealing with the two spatial extremes with which they are faced as new arrivals. On the one hand, they have to become accustomed to containing their daily lives within the limited confines of the bedroom (Peace et al., 2006), the only space which they can really 'colonize' with their own possessions and in which they have to forge a new identity. As one resident summarizing the feelings of many who found it hard to downsize and to crowd a lifetime of memories and possessions into one tiny room put it, 'If only I had a bigger room – I've always been used to bigger and better' (RI/9). On the other hand, they have to learn to negotiate the public spaces in the home which are likely to be far more spacious than they have been used to and full of hidden physical as well as social challenges and pitfalls. One resident, for example, who had become disabled with Parkinson's disease and a stroke, said that since moving into the home he had stayed most of the time in his room, taking all his meals there because he felt embarrassed by his disabled hand and arm. He admitted to feeling lonely as a result: 'on occasions you feel down, and left out of things' (RI/73). Another resident was embarrassed by her incontinence and similarly found being in public areas of the home difficult. Some homes had been lavishly refurbished making the use of public spaces socially daunting for some. For others, such improvements undoubtedly represented change for the better in terms of living and working conditions since the late 1950s. In one home, which we visited in 2005,

the sitting-room had just been decorated, new curtains (gold brocade) were hung and a new carpet laid. In the past year a new stand-aid hoist, a new double freezer and a new photo-copier/fax machine have been purchased. The manager's latest plans for improvements have just been passed. They include a planned extension of the dining room for the use of the residents for hobbies. ... At the same time the conservatory will be re-decorated, with new seating and carpet tiles.

(HR/10)

Other improvements presented physical challenges for residents. One home, for example, T-shaped and purpose-built in the 1950s, had a lift installed for the first time in the 1960s. This was centrally positioned, between the main lounge and the dining room and adjacent to the main staircase. In 2004, the lift was replaced by a new lift shaft, installed at one end of the ground floor corridor. The old lift shaft remains but is now redundant. The effect of this we observed, although not commented on by staff, was that a substantial proportion of residents had much further to travel between the downstairs communal areas and the upstairs bedrooms with the potential consequence of undermining their independence and autonomy. Another home, a privately owned Grade II listed converted Victorian rectory with exquisite ecclesiastical features, had had substantial sums of money spent on upgrading it, but we found, as Townsend did in the 1950s among many of the converted homes he visited, that the building remains essentially unsuitable for meeting the care and mobility needs of many frail older people. Furthermore, in some of the converted homes we visited, the division of large rooms into single bedrooms with en suite facilities sometimes resulted in awkwardly shaped rooms which were far from ideal living spaces.

External spaces

Townsend was struck by the 'frequent contrast between the beauty and order of surrounding lawns and gardens and the dinginess or bleakness of the premises' (1962, p. 118). He noticed that some matrons and owners prized the grounds so highly that it appeared that they prioritized well-kept gardens over the standards maintained inside their homes. In 1959, one of the home reports recorded that the gardens of the home were

> some of the most beautiful I have ever seen, with beautiful lawns and two small summer houses, borders of beautifully banked flowers ... the condition inside the house does not come remotely near the state of the gardens.
>
> (PTC, Box 36, HI 3)

Similarly, many of the homes visited in 2005–6 aimed to give a good impression to the outside world through careful attention to the external aspect and the upkeep of the gardens. One manager, when asked for her wish list for the home as a whole, rather surprisingly considering other more practical priorities, said that above all she would like to revive and rebuild the once fine fountain in the front of the home.

Public appearances aside, gardens had an important role in the lives of residents both in the 1950s and in the early 2000s although their role has changed somewhat. At both times, they were places of work as well as leisure, an issue we return to in Chapter 7. An important function of gardens in 1958–9 was the privacy they afforded residents who otherwise shared rooms with several people. Residents could entertain their visitors in the garden, and this seemed to be the norm for one home, where the owner said that visitors could be seen 'in the garden or in the dining room' (PTC, Box 36, HI 1). Gardens also offered a neutral space for men and women to meet in otherwise strictly sex-segregated homes.

In 2005–6 gardens were less essential as a private place to meet visitors or to mix with one another. Some of those we visited had developed quite specific roles not mentioned in the 1958–9 research. The garden of one faith-based home, for example, aimed to offer a religious experience to those residents who were able to walk or to move around the wheelchair-width paths. Furthermore, in more recent years a new role for gardens has been developed as specially designed spaces for those with dementia (Chalfont, 2005, 2007, 2008). Four of the 20 homes we visited had particular external areas for those with dementia, some in the form of an enclosed garden, patio or internal courtyard. In one of these, two residents diagnosed with dementia indicated that they liked the garden and the flowers: 'If you have friends, you sit with them or go to their room, or sit in the garden – there are plenty of flowers' (RI/27). Another, with mental health problems, was able to take advantage of the wildness, peace and quiet, and bird-watching opportunities afforded by the grounds of this rural home (RI/26). Larger gardens like this offered the chance to exercise. The four acres retained by Pine Grange, for example, were used by some energetic residents who took regular walks as part of their daily routine.

The significance of gardens for care home residents has been highlighted by several researchers (Percival, 2002; Parker et al., 2004; Chalfont, 2005, 2007, 2008; Chalfont and Rodiek, 2005; Bhatti, 2006; Manthorp, 2008) and our own findings support this. Bengtsson (2004) suggests that gardens need to be both physically and psychologically accessible to residents. Barriers to physical access include consideration of distances, slopes, benches, thresholds, doorsteps and kerbstones while psychological accessibility requires visibility of the garden from inside the home and a confidence that it is a safe environment to visit. We found that in some homes in 2005–6 the garden which looked so inviting, was in fact quite inaccessible for less able residents and remained for them, frustratingly, only a view. The garden of one, for example, was laid out on different levels, with steps up to the lawned areas. Only one section – a patio area with garden furniture – was accessible to all. A resident said how much she wished that there was an accessible garden. She could see the lawns and flower beds from many windows – including the dining room – but could not easily reach them. Hers was not a lone voice. The now much reduced 'beautiful' gardens of Cefn Bryn mentioned above were difficult to reach, having to be accessed through the front door secured by a keypad. A similar arrangement had existed in another home until the new dining conservatory was added which, as one resident commented, was a bonus because it allowed independent access to the garden via the ramp leading off it. The gardens of another, carefully designed for those with physical and sensory impairments through the inclusion of a sensory garden, raised flower beds, seats and gazebo, remained inaccessible due to the shortage of staff to help those with mobility problems. This lack of staff support affected the resident of another home who had recently lost her sight and needed help and reassurance:

> In the summer time I sit in the garden. I'd like to *feel* I could get out but I'm afraid of falling. Also, the girls in here haven't got time – so I can't say 'Would you like to help me down' in the summer. I like to sit in the summer house.
>
> (RI/53)

This home stood in approximately a quarter of an acre of land with trees, flowerbeds and summer house. There were bird feeders, bird baths, garden chairs and a large wooden swing chair in an enclosed and sheltered part of the gardens – so it offered a sensory experience for this particular resident when she was able to visit it.

On occasions it was clear that although the 2005–6 managers subscribed to the importance (and occasionally the grandness) of gardens, they were not able to see their vision through. They sometimes praised their gardens, implying that they were well used, when the opposite was the case. For example, in one home, the manager was keen to display the gardens and was proud of their facilities such as the summer house. However, this picture was somewhat deflated later by the volunteer on the reception desk who remarked to the researcher that it was 'such a shame' that the gardens and the summer house were never used. Other homes had gone to considerable lengths to make the garden a major facility, so that it was accessible, well-maintained and beautiful at all times of year. Even in colder seasons, if the sun shone, tables and chairs were arranged, shawls and hot cups of tea and coffee provided so that the residents could sit outside. In these cases, the garden became a feature of daily life for several residents.

As important as the garden was a room with a view. Lewis (1973) describes this 'watchful space' as offering diversion and entertainment. More recently Peace et al. (2006) have revealed the importance of views across parks and gardens for those living in sheltered housing who also indulged in bird watching and people watching. Views were important to residents in both periods, although the significant difference between then and now was the fact that many people in 1958–9 did not have a window of their own to look out of, or the windows were too high so that there was no view at all. The importance of a room with a view was also noted in 1958–9:

> I was able to see the two rooms of the residents whom I interviewed and these were beautifully appointed. One of them was large, warm and had a beautiful view of the open square and the traffic passing outside. This gave the chair-bound resident whom I saw a great deal of pleasure.
>
> (PTC, Box 36, HI 2)

In 2005–6, most bedrooms, but by no means all, had views for the residents to enjoy, and certainly many of the public rooms faced out onto large gardens. Weather watching from a bedroom window gave occupation to one resident, who kept a daily diary for one week, with regular observations on views from her window of the November weather, the time of year, and the temperature registered by her thermometer.

> Sun behind trees as it's quite low now – Sun into room now, nice and warm – Another beautiful clear blue sky warm sun and slight breeze – Frosty morning with sunshine to clear it. First time my thermometer has been below 70F! – Looks like a sunny day! – Clear blue sky – pressure up but temperature falling, I think – No clouds to be seen from my room – gentle breeze.
>
> (RD/13)

In some homes we visited, gardens were carefully planned so that residents with rooms overlooking them could indulge in bird or animal watching. Bird feeders were often placed strategically to enable spotting, and unusual birds visited the feeders: nuthatches, for example, were pointed out during an interview with one resident. Such opportunities to enjoy gardens and views were one way of making homes more 'homely'.

Creating a 'homely' environment

The imperative to create a homely environment, based on a domestic, familial model of home was apparent in the 1950s, particularly in the management of the smaller private and voluntary homes. In 1959, the owner of one small private home accommodating 13 residents, for example, commented, 'We like to be as much of a family as possible' and in a letter to the researcher, he wrote:

> It is primarily our own Home for ourselves and our children. We try to avoid both the name and the spirit of an institution, and to take in a number of frail or elderly folk much as we would take into the family our own uncles or aunts if, owing to old age, any of them needed to be cared for. In fact, one of our guests is one of my wife's uncles.
>
> (PTC, Box 36, HI 7)

Several of the managers in the 1950s drew on familial metaphors such as the matron who said 'we want to be – a happy family together' (PTC, Box 36, HI 2). Another told the researcher that if she was sitting in a resident's special chair she would give it up to that resident 'like when father comes home, you get out of father's chair', and that the last thing she did at night was to kiss them all goodnight (PTC, Box 36, HI 11). And a husband and wife managerial team of one private home liked to be called 'mum and dad' (PTC, Box 36, HI 5).

By the early twenty-first century, the ideology of home remained strong as reflected in the frequently repeated trope by managers: 'it's their home'. As Peace and Holland (2001) have observed, however, the ideal of home, not least due to size, is seemingly unattainable. There may be a decorative veneer of homeliness through, for example, the inclusion of potted plants, rugs, old-fashioned furniture, ornaments and pictures in the public spaces of the home, but the accompanying homely lifestyle is more elusive. Lundgren (2000) argues that the 'aesthetics of being home-like' are an expression of conventional ideas and tastes which are imposed upon residents and are therefore an expression of power. Our findings in the main supported this argument and often it was not just the aesthetics of the home that belonged to the manager but also the arrangement of spaces, the routine, programme, culture

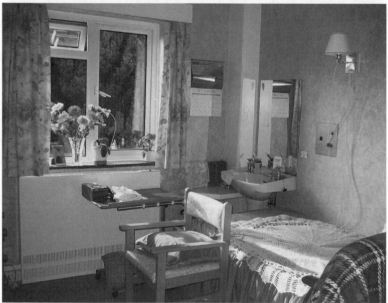

Photograph 6.3 Resident's bed-sitting room 1959 (above) and bedroom 2005 (below)

Source: The Peter Townsend Collection; *The Last Refuge* revisited collection. © Jean Corston (top), © Julia Johnson (bottom)

and ethos of the home, all of which could be equally controlling. As one resident so succinctly put it: 'We've had to mould ourselves into the kind of life thrust upon you – and it hurts' (RI/9).

In terms of the physical environment, there is no doubt that those homes which accommodated residents in large dormitories and shared rooms in the 1950s have become more homely through the increased provision of individual private spaces primarily in the form of single rooms. Nevertheless a comparison of bedrooms in the two periods produces some surprises. In 2005–6 we still came across single rooms in purpose-built homes occupied by residents we interviewed, which, although providing more privacy, were bare and uncarpeted, with few personal possessions in evidence. There were of course some residents who wished to keep a clear boundary between their own home and the residential home, perhaps in the hope that they would one day return to the former. One resident, for example, made clear that her furniture had remained in her own home, the implication being that that was where it belonged. We also found evidence of continuity in some of the converted homes. On a visit in 2006 to a small private home, for example, the researcher interviewed a resident in her shabby and unadorned shared double room which contained few personal possessions to brighten it and no chairs at all. Nearly 50 years earlier, the shabbiness of this home was reported 'especially in the bedrooms', with damp patches on the wallpaper and only 'adequate' furniture (PTC, Box 36, HI 11).

Several of the homes Townsend visited already accommodated residents in single rooms, however, and we have been able to compare his photographs of these rooms with ours. Among the collection of photographs taken in voluntary or local authority homes by Townsend are several which illustrate rooms similar to those in an ordinary family home. A photograph of a room in a voluntary home has attractive furnishings, large windows giving onto a view of trees, a comfortable armchair with cushions and antimacassar, a dressing table with personal objects, pictures and a radio (1962, Illustration 18). Another (1962, Illustration 21) is smaller and less elegant but very comfortable and homely, crowded with pictures, ornaments, a clock, a cup of tea, a calendar, a lamp – and the occupant is sitting on the bed doing some sewing. An unpublished photograph shows a shabby, untidy, but lived-in room with evidence of the occupant's occupations and interests. The top of the fireplace is crowded with ornaments, the pictures hang crookedly on the walls, but there are cushions and rugs on the three armchairs by the fire, a big kettle on the hob, a silver teapot on the table beside a sewing machine obviously much in use according to its prominent position in the room. Sun shines in from a window on the left of the photographer. These three rooms seem to validate the claims for a 'home from home' in the early period, at least in terms of personal space.

Several bedrooms we visited and photographed in 2005–6 could not have been mistaken for rooms in a family home. Instead they conveyed the impression of an institutional, even a hospital room, with a clinical rather than a homely atmosphere. Sometimes, the more homely touches, such as flowers, plants, photographs or fluffy toys were to be found cheek by jowl with commodes, large plastic waste bins or hospital style mobile bed-trays. In one home, two of the bedrooms were unashamedly 'hospital' style rooms, intended, seemingly for the ease of staff. In one, a doorless cupboard space was filled with incontinence pads in full view; while in another two bottles of liquid soap and a cleaning agent were on the table beside the TV, a plastic waste bag hung by the wash hand basin and incontinence pads were piled up underneath. The contrast with the personalized rooms photographed by Townsend could not have been greater, and indicate a very different approach by management, owners and inspectors in 2005–6 to conditions of daily living.

Other bedrooms observed in 2005–6, however, especially those full of personal possessions and furniture, or those in homes where a high level of untidiness was tolerated, came closer to a more familiar 'concept of home'. One new resident described her flatlet as 'home', although her diary entry places it in inverted commas. Whenever she described leaving it to walk down the corridor, visit the lounge or sit outside in the sun, she then wrote about her return: 'back in my "home"'; or sometimes, leaving out the inverted commas, 'visited K. on the way home'. In a reflective diary entry she wrote: 'Having my pictures up on the walls has really made my bed-sit feel so much more like HOME' (RD/13). This example also suggests, like Peace et al. (1982), that the 'flatlet' which provides enhanced personal space has the potential to meet the need for autonomy and control, and therefore comes nearer to homeliness than a mere bedroom.

Rules, regulations and routines

Despite best intentions, the ideal of the homely, familial setting, it has been argued, is often at odds with the reality of institutional living (Willcocks et al., 1987; Higgins, 1989). Furthermore, familial ideology, as we argued above, carries implications of control as well as care. Inevitably, both then and now, sometimes despite protestations to the contrary, there were rules and regulations which affected both working practices and residents' lives. In 2005, a resident with a diagnosis of dementia summed this up:

> Before, I thought this place was going to be a boarding school and this is what it has turned out to be! My last years at school prepared me for coming here. It is too closed in.
>
> (RI/27)

Others residents we interviewed felt that they did not have complete freedom to do as they wished, one saying that 'you have to use your common sense not to do the wrong thing ... you have to fit in with them. It's a bit like infants' (RI/42).

Notices

One of the most obvious manifestations of the existence of rules and regulations are notices. The matrons' ideal of overall homeliness was sabotaged in the late 1950s by the common practice, observed by Townsend, of the display of large notices (sometimes typewritten and pinned to doors or walls), particularly in the ex-PAIs, to regulate behaviour.

> [I]n dormitories, 'SMOKING STRICTLY PROHIBITED'; in w.c.s, 'NOW WASH YOUR HANDS', 'DO NOT SPIT – MATRON'; in washrooms, 'NO MATCHSTICKS OR RUBBISH SHOULD BE PUT IN THE HANDBASINS'; in dining-halls, 'VISITORS MUST NOT USE THE DINING HALLS'; in bathrooms, 'PATIENTS MUST NOT BE LEFT UNATTENDED IN THE BATHROOM'; and in corridors, 'NO UNAUTHORISED PERSON IS PERMITTED TO USE THESE LIFTS'.
>
> (Townsend, 1962, p. 92)

We too found notices displayed in all types of homes, most notably in hallways, entrances and reception areas, corridors, staff rooms, offices and kitchens. It would seem that there has in fact been a proliferation of signs and notices since the 1950s. This in part reflects increasing concerns and regulations relating to risk and the health and safety of staff and residents. There are also requirements for certain procedures to be on display so as to protect the rights of residents and relatives, for example, to make a complaint or be present during an inspection visit. 'Wayfinding' demands signage to ensure residents can move about the home easily and securely and are able to identify their own bedroom. Occasionally there were handwritten notices stuck haphazardly in the relevant spot: 'Do not leave the keys in the doors'; 'Heads or other limbs could get caught in the railings!!' And we came across bollards and signs forewarning passers-by of potential hazards and dangers such as wet floors. Some homes had displays of photographs demonstrating the involvement of residents in fund-raising and other organized activities. Residents appeared to find some notices useful, and in at least one home we observed them crowding round to read the list of church services and other activities for the week. However useful, such notices and displays inevitably contributed an institutional feel to the homes. Some managers recognized this, such as one who tried to present information in ways she considered attractive – a colourful board, coloured paper, illustrations – but who also remarked that she tried not to make the notices too intrusive: 'they are more for the staff than the residents' (MI/20).

Tobacco and alcohol

The claims by some managers in both the late 1950s and early 2000s that there were no rules for residents and that they were free to do as they liked did not always hold up to scrutiny. On the question of alcohol and tobacco consumption, for example, we found some scattered references in the 1958–9 reports on the surviving homes. Just one, a faith-based organization, banned smoking and drinking altogether. Prohibition of tobacco was unusual. Indeed, in the ex-PAIs and other local authority homes, the men were often given a weekly ration of cigarettes or tobacco, and the women sweets, on top of their personal allowances. In several cases, women were not given any form of allowance and in other institutions only sweets were provided. At Cefn Bryn, a former local authority home, the residents received '20 cigarettes or 1/3d worth of sweets' in addition to their allowance. Smoking was on occasions, however, restricted to certain areas. 'No Smoking Upstairs, By Order' (PTC, Box 36, HI 1). 'Smoking is not allowed in the dining room and main lounge' (ibid., HI 2). 'They can smoke where they like but they have to be careful' (ibid., HI 11). Townsend makes the point that there was considerable variation between homes. In his book, he describes a woman resident in one home 'sitting up in bed reading the *New Scientist*. She was smoking a cigarette in a gold cigarette-holder and her bed was littered with books and periodicals' (1962, p. 186).

By 2005–6, a White Paper on public health (DH, 2004) had been published. It proposed a smoking ban in almost all public places in England, a mere 50 years after Sir Richard Doll co-authored a paper that confirmed the link between smoking and lung cancer (Doll and Hill, 1954). Following contentious debate, a smoking ban in all enclosed public spaces and enclosed work places in England came into force in July 2007 with a number of exceptions, including residential care homes (Office of Public Sector Information, 2007). A similar ban had been introduced in Wales a little earlier in April 2007. The consequences of smoking were having an impact prior to the ban coming into force. Nine of the 20 homes we visited had a clear no smoking policy, though in three of them, residents could smoke in the garden. Eight had designated smoking areas within the home and three allowed smoking in individual rooms. However, the regulations were stricter for staff than for residents and all had to go outside the buildings to smoke.

Where alcohol was mentioned in the 1958–9 reports on the surviving homes, it tended to be in the context of control. In one home, for example, residents were allowed to keep and consume alcohol on the premises 'if the doctor approved' (PTC, Box 36, HI 10). In another, residents were given milk and brandy 'if they had chest trouble and the weather was bad' (ibid., HI 11). In a private home, alcohol was not allowed except 'at Christmas and then we supply it' (ibid., HI 1), and in another private home residents were permitted to have a cigarette and a spoonful of whisky in their milk, 'so long as we control it' (ibid., HI 11). It was also reported that one or two of the proprietors 'indulged rather freely in gin or whisky' or were 'on the bottle' (Townsend, 1962, p. 188, p. 202). Indicating some differences in the way men and women were treated with regard to alcohol, the men in one of the Catholic homes received four shillings a week (personal allowance) more than the women and were also given a glass of beer on weekdays and a glass of rhubarb wine on Sundays.

By 2005–6 the regimes on alcoholic drinks were relatively relaxed compared with the past. One resident for example, when asked what she did in the mornings, replied 'I dress, go out for a walk, have a little drink in my room – a glass of sherry – read the papers, tidy up my room' (RI/10). Nevertheless there was still a significant degree of control. Although, 15 of the managers we interviewed reported that residents kept alcohol in their rooms, seven of them said that a check was kept on the possibility of a clash with medication. The remaining five managers operated stronger controls. 'We like to keep it in our own cupboards and they can ask for it as and when required because some have been known to have one too many' (MI/18). 'There is a bar but the licence has lapsed. We normally look after their alcohol for them – it has to be locked up for obvious reasons' (MI/13). Unlike the late 1950s (Townsend, 1962, p. 166), five of the six voluntary sector homes run by religious bodies did permit alcohol, though one said that their policy had changed only recently. As in the past, however, the official prohibition remained in the sixth although it was accompanied by 'unadvertized flexibility' (MI/8). Alcohol was also described as featuring at weekends or on special occasions. Seven managers referred to Christmas, parties, special occasions, Saturday night trolley drinks and sherry or beer with Sunday lunch.

Overall, between the late 1950s and the early 2000s, there had been a sea change in the attitude to smoking, though in 2005–6 over half of the managers reflected a willingness to tolerate smoking within the limits of the law. However, the notion of the weekly allowance of tobacco, cigarettes or sweets had completely disappeared as a relic of the Poor Law days. Prohibition of alcoholic drinks had also disappeared, being replaced by some measure of risk assessment, whether associated with medication or possible overindulgence. The use of alcoholic beverages on special occasions was another more subtle form of control. In brief, there has been more change than continuity in the regulation of both tobacco and alcohol consumption, but with more control over the former and less for the latter.

Pets

Another issue which on closer inspection, at least in 2005–6, was clearly regulated related to pet ownership. Although generally speaking, residents are now more strongly encouraged than in the past to bring with them their own possessions and items of furniture, the line is often drawn at pets. There is no mention of pets in the 1958–9 reports although some communal pets, such as budgerigars, were in evidence in Townsend's photographs. In 2005–6, eight of the 20 managers we interviewed said residents' pets were not allowed, though one commented that a cat would be allowed, if requested, but not a dog, as there was no garden, and another stated that the 'no pets' policy was flexible, and that two residents currently had cats. In a third there was a 'communal' cat (and a goldfish tank). Fish and caged birds were mentioned as residents' pets by six managers, and one home that did not allow pets had accepted a donation of an aviary of cockatiels for the garden. Where pets were accepted, there were usually clear limitations laid down. Either 'appropriate accommodation' had to be available or limitations were placed on the kinds of pets allowed: caged birds were 'as far as we can go' (MI/2); 'we can't have extremes' (MI/15). Some managers said pets were allowed on an individual basis or that the decision would be determined by a risk assessment. One manager said: 'I honestly don't know. There are so many risks associated with various animals. Birds have psittacosis, cats have toxocara. There are so many rules and regulations. I think rabbits are the safest' (MI/13). Another manager worried that staff would end up looking after the pet. As far as cats and dogs were concerned, there were overall in the 20 homes we visited five 'communal' cats, two 'communal' dogs and visitors accompanied by a pet (usually a dog) were mentioned by three managers. Just one home appeared to take a particularly positive approach to pets.

> We've always had pets. Residents can have pets, but no birds because the manager is allergic. We have had guinea pigs, rabbits, three elderly cats, and now we have a house dog. ... We try to include people and ask their opinions. We discussed having the greyhound.
>
> (MI/20)

The overall impression given by the managers was of ambiguity at the policy level leading to great variation in practice. The reasons given by managers for the views they expressed focused more on the consequences for the running of the home rather than on the preferences of the residents. An inquiry into mental health and wellbeing in later life reported that having pets is one of the important elements promoting wellbeing (Age Concern and Mental Health Foundation, 2006) suggesting that ways need to be found of developing more pet-friendly policies in care homes. In 1998, it was estimated that each year, prior to entry to a care home, about 140,000 pets were handed over to vets or animal sanctuaries and 38,000 of them were destroyed (Anchor Housing Trust, 1998; Valios, 2009).

Routines

In both periods, daily life was dictated and punctuated by set routines. In 1958–9 it was reported that routines were on the whole strictly adhered to in the homes – the routine was 'fairly rigid' in particular in the large institutions. Bed times were fixed. In the ex-PAIs, 'a very large proportion of residents were in bed by 6 p.m.' and this was because of staff deadlines rather than residents' preferences (Townsend, 1962, p. 98). In private homes there were similar restrictions. There, the researcher was told, 'Residents must be in bed soon after 8 pm. We must turn out their fires. We haven't the staff to attend them in the evenings. We wouldn't have any life of our own if we couldn't reckon on sitting down by then' (ibid., p. 197). Mealtimes were not flexible. In one home 'Everybody gets breakfast in bed at 8 a.m. and lunch is at 12 noon. Afternoon tea is at 3.30 and the evening meal is at 6 p.m. Residents take this in their own rooms' (PTC, Box 36, HI 10). The researcher criticized one home in his report, saying that 'In retrospect, not a very good home. ... Matron selects the fittest applicants and then subjects them to a quasi-hospital regime' (PTC, Box 36, HI 11).

The most notable difference in the routines in 2005–6 was their stated flexibility. Every home we visited made a point of this, emphasizing that choice for residents had priority, and highlighting the desirability of a home-like atmosphere. There were occasions when, through our own observation, it appeared that the flexibility was more an intention than a reality, but on other occasions the wishes of residents were fulfilled. We found that the day started for people at different times. Usual getting-up times in one home ranged between 8 a.m. and 10 a.m. In another home, there were on the one hand early risers who were accommodated, and on the other those who wanted to stay in their rooms for breakfast or not to dress until much later in the morning. In one home, if a resident wanted to stay in bed the manager would agree, though she would also check for any signs of illness. Such behaviour would not have been tolerated by the manager in 1958 who felt that 'it would be the end' if she allowed people to stay in bed. We found that meal times were also said to be flexible, though we observed that people usually started congregating in the dining rooms at about 12.30 p.m. In one home high tea was a moveable feast, from 5 p.m. onwards; and late drinks were between 7 p.m. and 8 p.m. and again between 9.15 p.m. and 10 p.m., with opportunities for hot drinks and toast at times during the night for those who could not sleep.

The fixed weekly routine for bathing which existed in most homes in 1958–9 we might have expected to see discarded by 2005–6. However, in two of the homes we visited a bath rota had only just been challenged by new managers who were trying – against the opposition from more traditional staff – to replace it by personal choice for each resident. Managers adhered to the idea that bathing should be flexible. One home reported that baths were twice-weekly at least, but more frequent if possible or as requested. Interestingly, this home had a similar procedure when Townsend visited it and found that some residents bathed every day, and all of them at least once a week.

Apart from routines relating to the physical care of residents, there were also routines relating to their social needs. Townsend recommended that 'arrangements should be made to celebrate each resident's birthday' (1962, p. 423). In 1958–9 there were often no special arrangements for birthdays, with some homes arranging monthly parties or as Townsend was told in one home which accommodated 203 residents: 'We do 90s and upwards' (PTC, Box 36, HI 8). Clearly, Townsend's recommendations regarding birthdays have been taken up during the intervening years, often to the extent that their celebration had become something of a standardized routine. In one faith-based home individual birthdays were celebrated in 2005–6 with a thanksgiving service, and the residents could choose their own hymns for the service. They would then be given a card, a present and a birthday cake – and the other residents joined in singing 'Happy Birthday'. There was much celebrating in this home – wedding anniversaries were similarly marked, with a thanksgiving service, cake, party and friends invited into the home. Unusually in one home, we came across the annual celebration of each resident's arrival date at the home – the 'Coming Day' – celebrated with flowers and in another (women-only home) every resident received a Valentine's card and roses. Of course not everyone, as Bytheway (2005) has noted, is comfortable with age-based celebrations and the real question is the extent to which the homes we visited varied their responses in line with individual preferences. When asked the question, 'Are residents birthdays celebrated?' all the managers said 'yes'. All mentioned cakes and often the keeping of a list of birthdays in the kitchen. Several mentioned singing and partying and the involvement of families. A minority acknowledged that not everyone likes a fuss and it depended on the individual, and one noted that 'a lot don't join in'. The change in culture relating to birthdays in the early 2000s, with all kinds of anniversaries often being extravagantly celebrated seems, in part at least, a reflection of the general commercialization of such events. However, it is also a reflection of the fact that on average the homes were smaller than in the past affording more opportunity to meet individual needs and preferences.

Managerial styles and staff attitudes

The 1958–9 home reports contain very little qualitative data on the way the care attendants went about their work. They were not interviewed and so we too did not interview any of the care staff in the homes we visited. In consequence, their voices are not heard and, as mentioned in Chapter 3, our data contains little about the bodywork carried out in the homes, one of their main functions. In comparing the qualitative data on staff in the two time periods therefore, we must focus on the managers and management styles rather than the experience and day-to-day practice of care work.

Management styles were different in the two periods, the available evidence indicating a more authoritarian and sometimes unsympathetic approach in the homes when Townsend visited them. Then, some of the managers, when interviewed, sometimes expressed harsh attitudes towards residents. The matron of one of the private homes, for example, was reluctant to allow Townsend to interview the residents, and described one as 'a real old grumbler. She's always grumbling and I want to get rid of her. She'll tell you she wants a room of her own, but don't take any notice of her' (PTC, Box 36, HI 10). Townsend also noted that this matron showed little respect for privacy, never pausing to knock on the bed-sitting room doors and one resident, who was described as 'senile', was locked in her room. In another, a local authority home, the warden was heard 'shouting at one of the residents somewhere at the back of the house', and she was quoted in the home report as making several derogatory remarks about residents in her care:

> All they do is sit and jangle. Some are not bothered. ... You see, some of these are low-graded. ... They're not interested in anything. ... We have had some who wanted to go back even to the old workhouses. I think some revel in living in their own dirt. You've got to be on top of them.
>
> (PTC, Box 36, HI 10)

In 2005–6 none of the managers we interviewed were quite so outspoken suggesting considerable change. Even so we encountered some powerful figures with very strong ideas about the way their homes should be run. For example, as noted earlier, the manager of one home was so anxious to impose her own view of 'a community' on the home that some residents felt pressured into coming down to the lounge every day and joining in communal activities.

Although we were unable to observe what happened behind closed doors, interviews with residents revealed unsympathetic attitudes and unduly controlling practices in some homes. Staff were often described by the residents we interviewed as 'friendly but not friends'. Some hinted at a lack of understanding from certain members of staff responsible for their daily care such as one who said 'I'm not happy. The staff are terribly busy – they haven't time for TLC. Some are really caring, really kind – but they vary' adding:

> I've had some spats. ... It sorts itself out. I've not quarrelled really – but I'm very independent. Sometimes the staff are not very kind. I asked for a pad and she said 'Well, reach over and get it yourself.'
>
> (RI/2)

It is important to note that more sympathetic attitudes were reported in several of the homes visited and staff were often described as marvellous, and once as 'angels'. These more positive views were particularly the case in the voluntary homes, illustrating substantial continuities. One of the voluntary homes, for example, had a very favourable report from Townsend, who noted that 'the excellence of this home does not lie only in its high standard of amenities. In fact there are parts of this house that are decidedly shabby. ... The really pleasant thing about [it] is the way all the rooms look lived-in' – a fact he put down to the humane and civilized atmosphere of the home. This was especially noticeable, he said, when the matron 'is talking to or being talked to by one of the residents'. He described it as 'one of the pleasantest that I have visited' (PTC, Box 36, HI 2). When one of us visited this same home in 2006, a similar managerial style was observed, one that encouraged and facilitated the relaxed and sociable use of warm and comfortable public rooms. In one of them

> [o]ne resident was entertaining a visitor [...]; another was doing a puzzle and talking to his neighbour. People were talking to one another – some, I noticed, asking questions of one another to set their minds at rest concerning forgetfulness.
>
> (HR/9)

In 1958 it was noted that the matron objected to uniforms in residential homes and had ordered 'floral overalls' for her staff. In 2006, the manager had the same views about uniforms, saying that they were 'institutional', and she said she was disappointed that her staff did not agree with her. This home could be held to be an example of continuity between the two periods – both the reports noting the sympathetic atmosphere in the home and the warm, unpatronising relationship between the manager and the residents.

Conclusion

Townsend's theory of institutionalization was powerful and contributed to the drive to make homes, particularly publicly owned homes, 'more homely'. As this chapter demonstrates attempts to de-institutionalize the institution have included the replacement of the ex-PAIs with smaller purpose-built homes and the reduction in size of the other large homes. In addition, the harsh and sometimes punitive attitudes displayed by some of the staff Townsend met and the rigid rules and routines they imposed have mellowed over the years. In such respects, the quest to make residential care less institutional has been a success. At the same time, however, some of the homeliness and informality of the past has been lost, particularly in those private and voluntary homes which were relatively small. Several homes have increased in size and now employ far more staff looking after an overall more (though not entirely) frail population of residents. In all the homes we visited, the living environment, including staff practices, is now closely regulated through

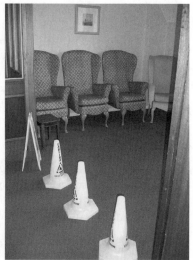
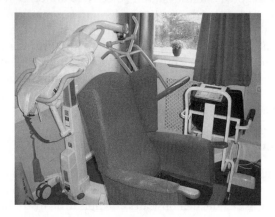

Photograph 6.4   The 'institutional' environment 2005–6
Source: The Last Refuge revisited collection. © Julia Johnson (top left), © Sheena Rolph (top right and bottom)

a wealth of legislation which in many ways constrains daily life for residents and compromises the ideals of choice and autonomy.

In discussing the attempts of matrons and wardens to create community life in the post-war local authority homes, Townsend was not unsympathetic:

> Fundamentally they were the victims of confused systems and policies. They were responsible for the welfare of the residents but,

at least to some extent, were supposed to respect the independence of individuals. On the one side they were attracted to concepts and practices associated with well-drilled hospital units. ... On the other side they were beginning to be aware of concepts associated with individuality and self-determination; with the warmth and security of home and family, and with the spontaneity, variegated disorders and cross-current loyalties of community life.

(Townsend, pp. 146–7)

At this point he addressed the fundamental question of how you make an institution homely:

> Do you behave like a hospital matron, a hotel manageress or a devoted daughter? The post-war Home was not a hospital and yet it was not a home. It lay uncertainly somewhere between and in a thousand details there was no clear answer to the problem of how it should be managed.

(Townsend, 1962, p. 147)

These words could well have been written in 2005–6. Townsend's conundrum about the problems of creating 'a home from home' remains. While owners have clearly implemented some radical physical changes in the fabric of the surviving homes in order to alter the living environment for the better, the managers retain responsibility for what goes on day to day. They continue to struggle to strike the right balance between rights and risks, and to resolve the contradictions inherent in the ideal of a well-managed home life.

What we have also drawn out in this chapter, through comparing our home reports with those written in 1958–9, are some of the continuities in the living environment. This is particularly the case in homes which have been subjected to fewest changes in terms of tenure, size and ownership. In these homes, we have observed the transmission of certain cultures and practices from one generation of staff and residents to the next. So, while in many respects the living environments have become more standardized, retaining institutional characteristics, these homes remain distinct through their strong sense of collective identity and unique histories.

7
Daily Lives

In this chapter, we turn our attention to the daily lives of residents both now and in the past. In comparing daily life in 2005–6 with Townsend's accounts of it, we have drawn on observations made during our visits, the interviews with residents and the diaries that some of them kept. In many respects, *The Last Refuge* painted a bleak picture of daily life in residential care in the 1950s and Townsend's photographs have left a lasting impression of loneliness, apathy and despair. Fifty years later, the image of care home life as one where residents are sitting around the walls, doing nothing, remains strong. But how accurate is this picture of life then and life now in the homes that have survived? To address this question, our focus in this chapter is on how residents spent their time during the day and what kinds of relationships they were engaged in both inside and outside the home.

Appendix 3 contains extracts from four diaries: two written by women residents in 1959 and 2006 respectively and two written by male residents in the same periods. They illustrate some of the similarities and differences between daily life now and daily life then. Although they were selected to represent a typical day in the life of a resident, it is debatable whether it is possible to identify one day in the life of one resident to make such a claim. Nevertheless these accounts are real – they are not averages – and a comparison is revealing. Turning first to the contrasts, the 1958–9 diaries indicate that some residents were involved in paid work which contributed to the running of the home. It also reveals certain freedoms, such as being able to go out into town, to shop unaccompanied and to smoke a cigarette. At the same time, however, daily life in these 1950s diaries was clearly lived in a much more public sphere and there was little privacy: not only were washing facilities and sleeping accommodation shared but much of the woman diarist's day was spent in a communal lounge, the only place to sit and chat with friends and visitors, to read, knit and watch TV. Life for the woman diarist in 2006 is more centred on the bedroom where she eats her breakfast, watches TV and listens to the radio, entertains visitors and talks to relatives and friends on the phone. Unlike her counterpart in 1959, she is very confined in her movements due to physical impairment which makes the view from her room much more important, as well as visits from friends and relatives. Life for the male diarist in 2006 is similarly more private but seemingly more empty. He makes no reference to friends or relatives and has little to occupy his time.

In both periods there were residents who were undermined by institutional living, and unhappy at the boredom and inaction imposed by the regimes of some homes, and the lack of meaningful occupation. According to Townsend, '42.4 per cent of the men, and 36.7 per cent of the women said they did not have enough to keep themselves occupied' (p. 342). Many of the residents Townsend met told him that they regretted the lack of work or useful activity and we found similar evidence of boredom and frustration in 2005–6. The comment of one resident interviewed in 1958 is echoed in that of a woman resident nearly 50 years later:

> It's a monotonous life, but you can't help it. After getting washed and dressed and having breakfast there's nothing to do but sit down. We sit and look at one another until it's time for dinner.
>
> (Townsend, 1962, p. 341)

> I stay in bed, get up, dust and polish, breakfast in the dining room, talk if people want to talk. It's morning, afternoon and evening and there you are.
>
> (RI/16)

From domestic work to daily living activities

As the early diary indicates, in 1958–9 there were still vestiges of workhouse practices in some homes, including the requirement that the residents contribute to the upkeep and management of the homes through different kinds of work. By the twenty-first century, opportunities for residents to engage in domestic work were limited and their lack has left a gap in the lives of some. Furthermore, such engagement was no longer conceptualized as work but rather as a form of activity.

Quite simply, in 2005–6, the homes no longer depended on the labour of the residents in the way that some large homes – and certainly homes with farms or market gardens – did in the late 1950s. Work encouraged by modern managers was therefore seen to be in the interests of the residents rather than of the economy of the home in general. The engagement of residents in domestic activities is now widely discouraged, however, because of safety and risk assessments. The manager of one home summed this up: 'they no longer lay tables because of hygiene restrictions; and in the past they did the washing up, but not now because of kitchen restrictions – health and safety' (MI/16).

Townsend argued first that loss of identity and the apathy that descended upon new residents were a result of lack of meaningful work or activity which suddenly stopped upon entry to the home and second that the minority who were given work were in fact 'supplementary staff' whose cheap or sometimes unrewarded labour was being exploited for the smooth running of the home. The proportions of residents involved in work were highest in the voluntary and post-war local authority homes and very few received any form of payment. None did any cooking although half had prepared their own meals when at home just before admission (Townsend, 1962, p. 339).

A large number of residents helped in St Peter's, then a large convent, and Townsend collected details of as many as 55 who did so without reward.

> There are 8 or 9 doing all the washing up, with the help of a Dishmaster. There are 3 laying tables, 2 helping with vegetables, 30 who do dusting and sweeping of one little room, and most of the residents make their own beds. One man helps with the stoking and 2 with the gardening, 2 or 3 run messages; 5 help with the sewing, and 2 with the laundry.
>
> (PTC, Box 36, HI 9)

A collection of photographs in the convent's archive (dating from a period possibly just before Townsend's visit) seems to corroborate this information and illustrates some of the important tasks undertaken by residents, including 'old men who worked as many hours in the grounds as employed gardeners' (1962, p. 126). One photograph shows a nun working with two of the residents in a dormitory. The nun is polishing the wooden floor to a high sheen while one female resident in an apron is arranging flowers, and another is making a bed. In another photograph, a nun and male residents are feeding chickens, while a third reveals the importance of the male residents in providing food for the home: three, accompanied by a nun, are digging the ground and planting potatoes in a large vegetable bed. The photograph of the sewing room includes a long table at which three female residents seem to be mending and folding items of laundry, with a nun overseeing them. Like the 1959 diarist, those helping out in Hillcrest, a former workhouse, received some reward:

> 11 men and 4 women help regularly and receive rewards averaging about 5/-. ... The duties include washing up, which is done by a woman. A man of 66 'spends a lot of time in the dining room' in the course of laying tables, etc., another man helps in the greenhouses, one acts as a stores porter, one runs messages and one does dustbins. 'The pig man gets 10/-. He cleans them out and feeds them on Saturdays and Sundays'.
>
> (PTC, Box 36, HI 8)

It was not only the larger homes which encouraged work. In one smaller voluntary home of 28 residents, four men and 12 women helped on a regular basis with a variety of domestic tasks, and there was 'no shortage of volunteers for cleaning bedrooms when needed' (PTC, Box 36, HI 2). Others helped with shopping and gardening, and 'the old able-bodied seaman voluntarily helps with the lavatories after a messy resident has used them' (ibid.). The matron told the researcher that her night duties were aided by one of the residents: 'Her last duties of the day are "seeing that all the windows and doors are shut, checking lights and looking in on some. A trusted resident does the top floor for me"' (ibid.). The general policy set out in this home was that residents were encouraged to help, and it was reported that 'most of them like to'. In fact, there was competition for jobs, and the older residents tended to be jealous of the younger residents who tried to 'muscle in on their jobs'. In another voluntary home it was one of the conditions of accommodation in 1958 that 'Residents will be asked to make their own beds and dust their rooms', although all other domestic jobs would be covered by staff, unless there was a shortage of staff, when residents might be called on to help (PTC, Box 36, HI 2). In a voluntary home for men only, the general policy was to encourage light duties: 'If a man wants to do something, then he can. We have no hard and fast rule' (PTC, Box 36, HI 1). The researcher found a relay of men who undertook useful jobs in this home, including washing up, peeling potatoes, looking after the laundry, cleaning the silver, bed-making, hair cutting, cleaning the brass and looking after the shop (ibid.). However, in one of the local authority homes, domestic chores

Photograph 7.1   Cleaning silver, 1959
Source: The Peter Townsend Collection. © Jean Corston

were seen as women's work: 'I get the females to do their own beds and rooms if they can but not the men' (PTC, Box 36, HI 3).

Expectations varied, however, and there prevailed among some the view that work was quite inappropriate – a reference to the 'hotel' model of care. In the private homes in particular, one owner told the researcher 'We don't expect it', another pointed out that residents pay and another that, when asked to help with the vegetables, 'they say, "Why should I do it when I pay?"' There were also proprietors who did not want to be seen to be exploiting a free or biddable workforce. One said that 'the senile mentals' used to be given the job of dusting their rooms and the lounge, but he stopped this practice when the other residents began to criticize this as exploitation. He added, however, that some of the residents helped voluntarily in the laying and clearing of tables 'casually and informally' (PTC, Box 36, HI 11).

In contrast, in the former workhouse, there was evidence of some bullying by staff:

It was also suggested that the residents empty their own chambers, for one old man I interviewed was very upset about this. He could not walk or even stand without using two sticks and described the difficult process of getting his trousers on, and told me that one of the staff insisted that he should carry his heavy chamber pot himself. '"That's your job" she said. But how could I carry a heavy thing like that?'

(PTC, Box 36, HI 8)

Allied to the bullying, work was also used as a form of punishment, especially with those with learning difficulties. As one warden told Townsend: 'And then there's my mentally defective woman. The only way I can punish her if she's naughty I give her more work' (Townsend, 1962, p. 133).

Despite these examples of work undertaken by residents, Townsend found that most of the 'active' people did no work. Many said that they were bored: 'I get terribly bored. I would like to do something rather than waste my time day after day.' 'I've done nothing here. Not even washed up a cup. I've never been asked. I get fed up. In my own place I'd be doing things' (p. 341).

These quotations might not have seemed out of place in 2005–6 when even fewer were involved in the kinds of domestic activities that in the past made a significant contribution to the running of the homes. Indeed over 20 years ago, Willcocks et al., noted that 'in many respects, the physiotherapy of everyday activity is lacking in residential settings' and that this was a matter which exercised both staff and policymakers (1987, p. 50–1). They described the introduction of small domestic tasks in the modern homes as 'only cosmetic or token'. There is no doubt that this still carries some truth. We were told, for example, of the one resident who worked in the garden and another who helped with the tea; of those who helped to fold napkins and assist with the coffees if the home was short-staffed, and one who regularly went round with a trolley selling sweets and tissues and delivering the post; of those who sometimes laid the tables for lunch. The task of changing the date on the wooden date board was carried out every day by one of the residents in one home; and a male resident in another home which catered mainly for those with dementia acted as a doorman, checking that the door was locked.

In some of the purpose-built, former local authority homes, however, we found a stronger culture of 'promoting independence' and residents were encouraged to make their own beds and tidy their rooms. In one, a limited form of group living had been created in an otherwise traditionally run home through the provision of two 'self-help' rooms consisting of small kitchen/dining rooms enabling particular groups of residents to make their own breakfast and eat their meals together. They cleaned the kitchenettes, washed the crockery and cleared the tables – and on occasions baked for themselves and others. In another, each of the four wings, two for dementia care and two for more able residents, contained bedrooms, lounge, dining area and kitchenette. Although the dementia care wings were fully supervised, on the other wings two residents laid tables, got their own breakfasts and could make a drink for themselves in the kitchenette. A purpose-built voluntary home included some flatlets containing a bed-sitting room and access to a separate individual kitchen thus affording greater opportunities for engagement in domestic activities. The occupant of one described these as 'Cooking, shopping, cleaning, washing, ironing – I use the laundry room and have my own iron and ironing board' (RI/13).

As in the 1950s, however, there were also homes in 2005–6, where engagement in ordinary daily living activities of this nature was not encouraged. For example, one of the local authority homes, purpose-built in the 1950s, had always operated a 'block treatment' model of care and here the culture of care was, and remains, protective rather than enabling. When asked in 2005–6 whether any of the residents tended plants, the manager replied 'oh no, staff usually do that', and yet one of the residents told the researcher that she would like to tend plants if she had any to tend. Similarly, the manager said that there used to be a resident who helped with the dusting, 'but we don't get them like that any more' (HR/5). Her words echoed almost exactly those of the warden nearly 50 years previously who told Townsend 'We haven't really got cases of people who are fit enough to do things' (PTC, Box 36, HI 3).

It was also the case in 2005–6 that, as in 1958–9, the ethos in some of the private homes we visited was one of service, with residents not expecting, or expected, to be part of the provision of that service.

Caring and supporting

Apart from domestic work, Townsend identified a second category of work, caring, and noted that among the residents 'You usually have a Florence Nightingale in every Home' (1962, p. 103). We found evidence of this caring work in some of the 1958–9 home reports. For example, one proprietor is quoted as saying 'The old people are very appreciative and delightful with one another. The more active help the crippled' (PTC, Box 36, HI 7) and in another private home it was reported that the residents helped one another, and the matron when she was ill (PTC, Box 36, HI 11).

We found quite widespread evidence of residents engaging in this kind of caring work and of the considerable satisfaction it evoked. Although we came across the occasional resident in 2005–6 whose efforts to help fellow residents had been rebuffed and who, as a consequence, had decided not to 'interfere' further and to leave responsibility for their care to staff, plenty of the 'Florence Nightingales' discovered by Townsend were still to be found. There was, for example, one such resident in Laburnum House whose role was to feed a resident with dementia who was unable to talk or move very much. She carried out this task at every meal – a great help to the staff – and rewarding for herself.

> I enjoy helping. I feed F – every day I feed her. She can't hold a cup. I feel I'm doing someone a favour – I feel I'm helping. She loves it – she laughs.
>
> (RI/50)

In several homes the more able residents took on the role of shopping and posting letters for those who were unable to leave the home. One manager noted those who visited other residents sat with them, talked and read to them. Some helped one another in very practical ways, pushing wheelchairs or guiding those lost down the long corridors. As one such guide commented 'We often do things for each other' (RI/39). In the faith-based homes, residents who were retired priests, vicars or ministers said daily Mass or took other religious services, while one 94-year-old resident played the piano for the services, and others arranged the lounges for the services, setting up the lectern, checking the microphone and providing a glass of water. It is worth noting that the life of the home would have been considerably poorer without the services of the priests and pianists – in fact, it may be the case that sung services could not have taken place with any frequency without them. One retired priest commented:

> Though not officially appointed, I am a kind of chaplain, being the most able bodied of all the residents and much younger than most. ... I say the Last Rites for dying people. I say Mass every day at 9.30 am with another priest.
>
> (RI/20)

He was not the only unofficial appointee. A retired music teacher, who had named herself 'resident pianist', gave regular piano recitals as well as playing twice weekly for church services in the home. She observed 'I help them through my music – I'm certain of that.' Indicating an element of reciprocity, she added that other residents helped her 'by being interested and enthusiastic about what I am doing – one lady enjoys certain pieces of music' (RI/48).

Not mentioned by Townsend, although no doubt a typical feature of institutions in the past, were those residents (usually male) who took on the role of entertainer and generally keeping people's spirits up. One man, aged 101, entertained the 'ladies' and staff with his singing and joking: 'I help the staff because I joke with them, make them laugh and make them feel comfortable' (RI/44). Another, aged 61, not only ran errands for fellow residents but also made similar comments about how he kept up staff morale by bantering with them. These two men, of very different ages, also exemplify the desire among many residents we interviewed to avoid the stigmatized role of 'resident' which is associated with passivity. It was by finding a particular role, which made them special and different in some way, that they were able to maintain their dignity and self-respect.

I try to help the girls and I always strip my bed – my sheets and pil- lowcases. … I feed the dog and I keep the receipt for the dog food. I take the dog out for a walk with R [Manager]. I feed the birds – and then I ask if I can help with anything.

(RI/50)

While one of us was interviewing the manager, this resident knocked on the door of the office, having taken delivery of some dog food and other stores, in order to hand over the receipt.

Organized group activities

The declining opportunities to engage in ordinary domestic activities in the homes in 2005–6 may account for the increase, since 1958–9, in organized recreational activities. In 1958–9 little emphasis was given to such activities by the matrons or wardens who were interviewed and Townsend himself was equivocal. As he put it, 'Do you submit them [the residents] to entertainments, religious services and occupa- tional therapy or do you leave them to decide their own activities?' (Townsend, 1962, p. 147). Some wardens and matrons in the late 1950s eschewed organized activities on the grounds that they did not conform to 'ordinary' home life. In one small private home, for example, the main interest of the residents, according to the proprietor, was his fam- ily: 'Our old people love our young people. Babies delight them. They really enter into their lives' (PTC, Box 36, HI 7). This domestic scene, in his view, obviated the need for entertainment or other organized activities. In other homes, residents were reported to find such activities redolent of institutions and therefore unwelcome: 'The old people don't like things arranged. If there are organized entertainments they feel we are treating them as a charity' (PTC, Box 36, HI 2).

Not all owners and managers in the late 1950s were averse to organized activities, such as whist drives, snooker, film shows and walking, visits from the local Operatic Society, a small winter card school and a magic lantern show at Christmas, dominoes, crib schools, games and billiards. At Fairview there was an occupational therapy workshop where five men made stools worth £250 a year. The occupational therapist visited every three weeks and had originally been part of the staff of the home. Apathy regarding arranged entertainment was widely reported, however, and in another home it was reported that: 'Concert parties have got disillusioned because the residents don't attend. The local amateurs used to come and give a sing-song, but they didn't get enough there' (PTC, Box 36, HI 3).

In 2005–6, far more attention was being paid to organized activities programmes, though their quality varied greatly. We visited only one home, a convent, which, as in the past, did not organize any activities or outings on the grounds that there was no interest among the residents. Despite the fact that residents were involved in religious activities (the home had its own chapel), it was heavily criticized by the CSCI inspector for adopting this position. Organized activities in the other homes we visited in 2005–6 tended to follow a common pattern which included quizzes, bingo, sing-alongs, games, exercises of various kinds (such as chairobics) and craftwork. In addition, some homes had aromatherapy, reflexology and fingernail painting sessions. Activities programmes were often posted up on a wall. In one home they were advertised on white boards and took place in different parts of the home each week, for example, 'Board Games and pop corn on Malvern', 'Film and popcorn on Cheviot' and 'Board games and making paper chains on Nevis'. Other activities organized in the homes – such as art classes, woodwork, reading groups and regular recitals – appealed to professional and middle-class residents. A photograph display in one private home depicted a 'war memories' day when children from the local school had come to hear residents talking about their experiences. This home also had regular visits each month from a group of ex-servicemen who came to talk with residents. The weekly visit from a pianist was paid for by the amenities fund but, due to its popularity, the residents sometimes paid for extra recitals. Many of the homes, like this one, had regular fund-raising events to support group activities and outings.

Overall, however, organized activities were viewed differently by different residents. For example, in a local authority home one of the residents said she loved the occasional half-hour activity sessions, which included a quiz, bingo, sing-alongs, exercises, making hanging baskets, Easter bonnets and Christmas cards, and that they 'had a good laugh', but another was not at all keen to join in any of these activities. The simple message seems to be that one size does not fit all. What felt institutional for some provided sociability for others. Clearly there was also the potential for some residents to feel infantilized by some of the activities on offer in some of the homes such as children's colouring books provided during 'Painting', balloon bouncing, bubbles, and play dough and large floor dominoes and other board games. As one resident commented 'I don't play snakes and ladders – I feel I'm past that!' (RI/35). Others, however, enjoyed these kinds of activities.

> I like playing games. We like to have a big blue and white ball and throw it to one another for exercise. We throw rings and quoits. I did colouring in this morning.
>
> (RI/37)

Managers admitted that it was difficult to provide appropriate activities for all at the same time, with the result that a lot of the games were aimed at those who were less able physically, or who had a degree of dementia. The greater number of residents with dementia is one of the greatest differences with the 1950s. One manager of a home caring for 26 people diagnosed with dementia, had invested in Bradford University's *Dementia Care Mapping*, and said he found it a useful tool. Inspired by the ideas in the pack, he defined the aims of individual and group activities in the home as the encouragement and enabling of interactions between staff and residents, and between residents, together with the creation of safe spaces for people to walk around in (repetitively if they wished).

Most homes enabled freedom of choice in regard to joining in activities, and it was clear from our interviews with residents that some, having reached this time in their lives, wanted peace and quiet and preferred to remain in their rooms, or to venture out only occasionally.

> I sleep, sit in a chair and dream. I'm quite happy. There is not much I can do. I'm often tired and don't want to do anything. I am content to sit here. My mind is full of other things – family affairs.
>
> (RI/21)

There was one home, however, where the stated aim was to create a thriving Christian community, and residents were expected to join in bible reading, prayers, hymn singing and a daily service together with a varied programme of activities throughout the year which included a book stall, a clothes shop, singing in the afternoons, slide shows, barbecues, pottery classes, games and quizzes – with visitors and volunteers helping with special events such as firework displays and a carol service. One resident felt pressurized to join in.

> I wouldn't say we are free to do what we like. They like you to go down into the lounge – I think they call it socializing – but some are not well and you can't talk to them very well. We've got to abide by. ... If I wanted to stay in my room ... they prefer me to be downstairs. We've got to socialize. I like to stay in my room. ... They say I mustn't stay in my room because it's not socializing. So you can't stay in your room always.
>
> (RI/37)

In 2005–6, only one manager, compared to several in the late 1950s, queried the appropriateness of organized activities.

> We have two group activities a day. But we don't want to be institutionalized. We have bingo, old time music, the staff do reminiscence and current affairs. But it is not structured or formalized – not institutionalized.
>
> (MI/4)

During our visit to this home, a new batch of puzzles from well-wishers had arrived and the lounge was full of enthusiasts. Each resident who wished, had a puzzle on a tray in front of them – and the staff could not resist joining in. It all looked fun, a bit chaotic, and, the manager hoped, 'homely rather than institutional'. His reference to reminiscence raises another issue, however, and that is whether a poorly paid, educated and trained care staff can be sufficiently knowledgeable and skilled to undertake this kind of activity. As Bornat has observed,

> this side of recall is rarely studied and, though the practice is encouraged and well-embedded into activities programmes, often routinized in the same way as bingo and dominoes, the skills associated with engaging and supporting people in remembering are still not part of any officially recognized training, in the UK at least.
>
> (Bornat, 2010)

In a few of the homes we visited, the enterprise, skill and experience of residents themselves were drawn upon in arranging activities sessions. For example, one resident regularly organized slide shows. Overall, however, in 2005–6, activities programmes were put together and managed by staff. We found that it was at a more individual level that the enterprise and skills of residents were exercised.

Hobbies and pastimes

Townsend found that only 53 per cent of residents said that they pursued a hobby. Two-fifths mentioned knitting, sewing or embroidery and the others (divided largely on gender lines) named football pools, racehorse betting and playing cards. The researchers noted unusual hobbies and activities: one resident was learning to transcribe Braille into Tamil while another visited old churches. Others read novels or wrote letters, and some enjoyed handicrafts, gardening or woodwork. However, Townsend suspected that some of the residents claimed to do more than they actually did. He found that 'half of the remaining 47 per cent said that they only listened to the radio, watched television or looked at newspapers, and the others that they did nothing at all' (1962, p. 340). He put this inactivity down to apathy caused by the managers' low expectations of the residents and institutional regimes which did not offer opportunities or choices.

Interestingly, Townsend did not classify activities such as listening to the radio, watching TV or reading newspapers as 'hobbies'. It is perhaps not surprising, therefore, that the matron of one home claimed that 'very few people do anything' (PTC, Box 36, HI 10). The fact that residents were reported as reading a lot and talking to each other appeared not to count for much in the 1950s. 'Hardly any hobbies' were found in another private home, where the residents were said by the proprietress to have 'only three things to do – eat, sleep and natter' (PTC, Box 36, HI 11). Furthermore, there was some evidence that wardens and matrons viewed 'hobbies' as a female province. The superintendent of one large ex-PAI commented that hobbies were not taken up by men, whereas the women had knitting and sewing. Instead, the men played cards and dominoes, billiards and snooker, and did football pools (though the women had also organized a football sweep). Men were again said not to have hobbies in another home, unlike ten of the women including 'one old lady [who had] her own piano which she brought with her from Norfolk' (PTC, Box 36, HI 10). The men were reported by several managers as doing a lot of sitting outside, and this is attested to by several of Townsend's photographs. Phillipson et al. (2001, p. 9) describe this apparent gender difference as a feature of the 1950s, highlighted both in the urban sociology and the photographs of the time which depict older men as socially redundant and marginalized, particularly in working-class areas. Whereas the older women in these photographs are portrayed as busy, sociable and companionable, the older men are 'huddled on street corners, expressing a loss of purpose and identity' (ibid., p. 10). The managers Townsend met, who described the men's activity – or non-activity – as 'doing a lot of sitting outside' seemed to echo this picture. Likewise, in some of Townsend's photographs it is the apathy of the men that is featured, slumped or sitting in day rooms with little to occupy them.

There is evidence in Townsend's home reports, however, that in some homes residents organized their own entertainments, arranging tea parties in their rooms and going out to visit relatives: 'A few play cards and they invite one another into their rooms' (PTC, Box 36, HI 2). And there was a thriving social life in another home, with people dropping in to one another's bed-sitters 'all the time' (PTC, Box 36, HI 2). According to one manager, 'no-one required encouragement with hobbies as "They do it quite happily – turning their drawers out". She said that they all did something for at least two hours a week – "All read and write letters –10 knitters and fancy work"' (PTC, Box 36, HI 11).

In 2005–6 we still found some managers who commented on the apathy of the residents, saying, for example, that they could be more active but chose not to be. As in 1958–9, in 2005–6 the most frequently reported activity was watching TV. In some homes there were, however, residents engaged in some very creative and challenging activities, including quilt making, tapestry, model-engine making, rug making, music making, using the computer and writing – one poet had recently had her poems published. Some, as we mentioned in Chapter 6, were keen bird watchers. Unusually, one resident joined some training courses alongside the staff, gaining two diplomas on 'Dementia' and 'Vulnerable Adults'. None of these were activities mentioned by Townsend. However, other hobbies had much in common with those pursued in the 1950s. A prolific knitter showed one of us a drawer full of jerseys and cardigans she had made, and asked advice about a knitting pattern she was presently working with. Another, despite her poor and failing eyesight, crocheted tea cosies and offered to make one for the researcher as a Christmas present. Four musicians were able to carry on their music, two of whom had brought with them their own pianos. They not only played for church services but also entertained other residents. One man played the organ and studied it seriously, taking lessons. Another, a retired music teacher, practised on her piano every day:

> I am a professional musician – My Bechstein piano is here in an upper lounge – 10 am: Practise until 11 am – I have given mini Piano Recitals to the residents. They appreciate live piano music – 6.30 pm: to my Bechstein piano for playing until 7.15 pm – work on my Bechstein piano – Brahms and Mozart – Working on Mozart Sonata K280 [...] Am listening to Elgar Piano Concerto Radio 3 – Lovely, big sweeping tunes in Rachmaninoff's Second Symphony – Music is a wonderful therapy.
>
> (RD/48)

When interviewed she said that she felt that she had 'just enough' to keep her occupied: 'I've been a busy person all my life and would not like it to be less' (RI/48). Other residents reminisced about former hobbies and interests and some managed to continue to pursue them to a degree. One was a keen philatelist and was interested in cartography. 'I love reading maps. ... I had a car. We drove down country lanes with

Photograph 7.2   Recreation in 2006
Source: *The Last Refuge* revisited collection. © Sheena Rolph

my maps. Ordnance survey maps – up to Box Hill, high places, up in the Downs' (RI/62). He had brought both stamp and map collections into the home with him and he showed the researcher these together with books containing his own drawings done many years before.

As in the 1950s, we also found the occasional groups which formed to knit, play scrabble, or even sing together in one another's rooms. Passing time with friends was a significant activity for many and those who were not able to be so active or so independent appreciated the company of the other residents. One man noted as one of his main activities 'Being with the residents. Talking to them or just being with them' (RI/46).

Friends, cliques and internal communities

As regards the possibility of friendships within the homes, Townsend found that, despite the sudden communal life to which they had been subjected, new relationships were often tenuous (1962, p. 347). Some managers, however, claimed that friendships were easily formed and could thrive: 'Oh they soon pal up and then we try to move people around so that they can share a room.' Another spoke of the 'strong matiness between the men', with small card groups and schools being set up. Townsend noticed that some of the residents in the local authority homes made up 'informal groups or cliques', though they remained comparatively few in number, and close friendships were few (pp. 142–3). He made the point that 'it is often assumed that when thirty or forty old people eat and live together under a single roof they belong to a real community' (p. 143), but he found little evidence of this. In contrast, for those who lived in voluntary homes, and in single rooms, 'community life was often stronger. Friendships formed with other residents were more common and informal groups or cliques more numerous' (p. 169).

We also found in 2005–6 that cliques or groups occasionally formed within homes, and in such cases, daily lives were led to a certain extent within these smaller 'communities'. The more able residents who occupied the top floor of one home, for example, as mentioned in Chapter 6 had their own lounge, kitchenette and dining room and shared daily routines avoiding contact with the more disabled residents, particularly those with dementia.

> They [staff] did take me down [to the ground floor lounge], but it did distress me terribly. But there is a room upstairs for compos mentis people. I met S. who was lovely.
>
> (RI/2)

This group not only disassociated itself from other residents in the home, it also expressed dissatisfaction with the food and was working out ways to order a Thai meal occasionally which would be brought by taxi. One of the group already had her friends and her son calling in with pizzas and 'modern' foods such as muesli, couscous, peppers, watercress soup and aubergines which she had no hope of receiving in the home, despite numerous requests.

In one of the four self-contained wings of one of the purpose-built homes, a resident described how she was

> friendly with all, especially those in my lounge – they are all nice and we get on well together ... we play scrabble together ... we're all friends together.
>
> (RI/57)

She had been lonely in her own home and wanted to come into this home for the company. When we visited it, several people were sitting in 'her' lounge, which was pleasantly untidy with papers and magazines on tables in front of chairs, which were not arranged around the walls. The atmosphere was friendly with people telling jokes, laughing and talking. Several seemed busy with knitting, needlework or crosswords, and the manager commented that 'knitting is very popular among the women in this lounge, who knit together'. This group of residents seemed to know one another well and spent much of their time in each others' company, eating together in their small dining room and sharing the kitchenette.

Residents from some other homes said that their special friends were those at their table in the dining room, possibly promoting cliques that would be difficult to break into.

> Our table consists of four small ones put together and we are quite a cheerful group of oldies. ...
>
> Lunch – fish and chips and peas. Fruit cocktail. A lot of chat and laughter at our table as usual.
>
> (RD/17)

A resident in another home, when asked about her friends, mentioned the four people she sat next to in the lounge, played cards with in the dining room, and whose rooms were next door to one another: 'There are four of us on this corridor. Nice!' (RI/51).

There were also groups of friends who had shared identities, such as the six Polish residents who lived in a small unit with four others in a former local authority home recently transferred to the voluntary sector. They had their own lounge with Polish satellite TV and a dresser with Polish dolls in national costume and a crucifix displayed on it. Polish festivals were celebrated in the home, and information was translated into Polish. Volunteers from the local Polish community visited and arranged activities, and visiting priests said Mass in Polish. Other voluntary sector homes catered for particular communities of interest and contained small groups of residents who had known each other before coming to live in the home. For example, in some faith-based homes, there were those who had been members of the same local parish; and in the secular homes, those who had shared their working lives or interests. Our observations supported the findings of other researchers who have argued that a common culture or shared beliefs within specialist homes eases the formation of friendships (Reed and Roskell Payton, 1995; Kellaher, 2000).

Getting out and about

Individual outings

In the late 1950s very substantial numbers of residents were reported to go out daily from the homes into the local town or village, to take long walks, to go to church or to football matches. Indeed, residents were encouraged to leave the home as summed up by one matron quoted as saying, 'Two dozen go out every day so that they don't clamour for entertainment here' (PTC, Box 36, HI 3). This was particularly the case in the ex-PAIs and local authority homes where there were greater proportions of men. The superintendent of Hillcrest, for example, estimated that only 20 out of 203 residents did not go out nearly every day. As with hobbies, a gender difference was observed by the researcher:

> There were a number of men to be seen sitting in the garden and at the front gate and at the roundabout 75 yards down the road, and struggling up the hill back to the home. The women, however, seemed to sit indoors or, if out, to be going on some specific errand ... they like to go down into the town, which is a mile away. ... Especially on market days they no doubt find the town a much more interesting place than the home.
>
> (PTC, Box 36, HI 8)

Overall, fewer residents in 2005–6 were able to act independently and, in consequence, now had less freedom of movement. For example, whereas all but two of the residents in one home went out for a daily walk in the late 1950s, in 2006, none was deemed able to go out unaided, even with help. One never left the home at all and only one went out weekly with relatives. This caused great disappointment for two interviewees. One, who had some problems with her memory and had lived just a short distance away, found that she was not allowed to go for walks around her old haunts:

> I feel restricted ... I feel I'd like to go out on my own. I can go out with someone, but it's not the same. ... There's not enough staff.
>
> (RI/5)

The other, who had also lived close by was similarly constrained:

> I thought I could go for a walk along the cliff. But I can't go out unaccompanied. ... I'd like to have my own independence. I can't even go out for a paddle. ... It's a shame – for insurance purposes.
>
> (RI/8)

Like several people in the homes we visited, these two residents had chosen to move to a home in an area with which they were familiar – what Peace et al. (2006) refer to as a 'congruent location' where they could retain links with the past (Reed et al., 1998). Their feelings of confinement were exacerbated by the fact that the home had no garden, just a patio.

There were still homes, however, where the few who could go out without an escort did so several times a week.

> Have a Mozart Festival Concert in the Guildhall ... at 1pm. Walked up to the Guildhall where the Concert took place – a Trio of players. Clarinet, cello and piano. Played two Trios op.11 Beethoven Clarinet Trio and a Clarinet Trio by Brahms. Lovely.
>
> (RD/48)

A few days later she took a taxi to another venue to listen to a recital by two pianists. Whereas in 1958–9 the only way to move outside the homes independently was to walk and to use public transport, in 2005–6 other modes of transport were also available. A taxi took one resident to an embroidery group once a week and another resident had a motorized buggy:

> It's made such a difference having my 'scooter' – also with a lovely garage to keep the batteries topped up – out again on buggy to Chemist via Tesco, much of the way on the road, pavements are bumpy and not flat.
>
> (RD/13)

Daily walks were noted by another resident who was already planning for future walks 'when the weather improved': on day one of her diary she noted a 20-minute walk to the bus stop, a bus journey and a walk around the book shops; on day two she undertook the same journey and walked along the pier; on day three she went out with friends, catching several buses and walking; on day four she walked to the lake to see if the swans were back and to feed the ducks. On day five she went to the tourist information shop and got a new book and walked around WH Smith, bought some chocolate and returned 'home' by bus; on days six and seven, she likewise caught the bus and did some shopping.

A retired priest still had his own car which he drove to go shopping, to visit a friend in hospital, to give communion to other friends and to go swimming in the local leisure centre. And a woman whose husband was in a different home, which specialized in dementia care, drove to see him every day.

Of course it was a tiny minority who, in 2005–6, were able to enjoy such freedoms. For the vast majority, it would appear that the combined impact of increased frailty, shortage of staff, health and safety regulations, risk assessments and insurance anxieties, was that daily life for residents had over time become more restricted to the confines of the home. The majority, therefore were dependent on others to get out of the home.

Organized outings

In keeping with the more laissez-faire attitude of managers, organized outings were not common in the late 1950s. Although some homes organized annual – sometimes thrice-annual – outings, staff claimed that people 'could not be bothered' to turn up. Twenty people out of 203 at Hillcrest went 'fairly regularly' to the cinema in the town, but as regards organized outings in this home, there was only one a year. Presumed apathy on the part of the residents, expense and disinterest by wardens, all played a part in the rarity of such outings.

By the time of our research, a fair proportion of managers and owners of homes felt it incumbent on them to provide outings at various times of the year. They were potentially important in those homes sited some distance from local transport and other amenities and for those residents who had few visitors able to take them out or who could only manage in specially adapted transport. Some homes had their own minibuses. One voluntary home had two and trips were organized on a rotation basis for residents in different units – every week in the summer. There were also weekly outings: on Fridays some of the men were taken to the pub and the women to have coffee in town; on Mondays about half a dozen residents attended a local day centre. A local authority home also had access to the council's minibuses and in the summer months there were trips out to nearby towns as well as a Christmas outing. Another voluntary home was well placed in the centre of a rural town, with the residents able to enjoy the facilities of a beautiful park just across the road where bands played, artists showed their work and exhibitions were held in the tea-rooms. This home also organized well-dressing visits and other outings in the surrounding countryside. Four outings a year were arranged in another voluntary home, paid for by the very active League of Friends.

In some of the private homes, however, there were no outings. As one resident pointed out: 'This place has been sold. We used to have lovely trips into the country. I miss the trips we used to have' (RI/41). Unaware of this change, a recent arrival to the home suggested that 'external outings' would be an improvement:

> I think it would be nice if in the summer they could organize a coach with accessible seating for a drive around the ... countryside or to the seaside – to get us out of these four walls.
>
> (RI/39)

Another privately owned home likewise had no outings on the grounds that the residents were of such a high average age that they did not want to go out. The manager said that they had tried it in the past and people had dropped out at the last minute. Echoing the words of staff in the 1950s, several managers made similar claims: 'we are always fighting apathy – they'd rather stay put' (MI/11); 'they become set in their ways' (MI/9). Only two residents had turned up for a trip to the pantomime and several residents backed out of a trip to an old-time musical put on by a local school. It is difficult not to conclude an element of victim blaming in these instances. Some homes, however, were trying to tailor outings to individual needs. One, for example, as part of its 'person-centred' approach to dementia care, took only two or three residents out at a time to the pub, a garden centre or an area of the country with familiar memories. A voluntary home had similar plans for 'one-to-one' trips of a person's choosing and, in addition, if a resident was being transported for a hospital appointment, would invite another resident along for the ride.

The reluctance of several of the private sector homes to arrange outings, in contrast to the local authority and voluntary homes, suggests an unwillingness to devote extra resources to these events, particularly transport and staff time. For-profit businesses (as opposed to not-for-profit) may have difficulty in attracting 'charitable' donations, such as minibuses, or the time of volunteers. Given the age and frailty of many residents, there is a real possibility of last-minute cancellations which threaten the financial viability of pre-arranged trips. The consequences of this are serious. We were somewhat surprised to find, for example, that in one private home – a former rectory, immediately adjacent to the churchyard – residents were never escorted to services in the church. Rather, services were provided within the home. This tendency to bring services and other activities in rather than take individuals out was far from uncommon. Townsend would no doubt have seen this as a further step contributing to the institutionalization of residents.

The increased confinement of the majority of residents meant that, in 2005–6, visits from family and friends were particularly important and for many, they were the only means by which they could get out of the home.

The role of family and friends

Common experiences of loss and change meant that residents both then and now sought similar means to transcend barriers in order to alleviate their feelings of isolation and sadness. Townsend found that visitors were important, particularly for those with family. Overall, two-thirds of the men and nearly four-fifths of the women had been visited by relatives since entering the home (Townsend, 1962, p. 342). Only one-fifth of the residents were being visited more than once a week, however, and this seems to have been because some people had few relatives and a large number had no surviving children. More women than men, and more middle than working-class residents were visited – though not more frequently.

As mentioned Chapter 6, in 1958–9, many homes did not have a private space for families to visit – rooms were frequently for three or four people, and lounges were occupied. Establishing a new kind of relationship with families was therefore a problem for recently arrived residents, unused to the barriers – both physical and psychological – now placed between them and the outside world. One manager said that residents sometimes complained about having to use a general room for their visitors: 'It appears that when they see their relatives in theirprivate rooms they occasionally upset the other room sharer. ... But if you take a shared room you've got to put up with it' (PTC, Box 36, HI 10). Another significant finding related to shame and sadness at the circumstances in which residents found themselves. Townsend found that some residents recognized their loss of status and independence and said they were embarrassed to be visited: 'I'd rather my nephew not see what I've come to' (Townsend, 1962, p. 345). Instead, some residents preferred to go out to visit relatives: 'I'd rather go and see them. I'd rather go to their homes. You see a little bit of home life' (ibid., p. 345). One woman went to see her daughter who lived in a nearby street, and then spent part of every day with her. Twenty-one per cent of new residents had made such visits since they had come to live at the home. A blurring of boundaries thus took place between home and community – a blurring which made life in the home more tolerable and maintained links with the outside world.

In 2005–6, family and friends continued to play an important role in blurring the boundary between the home and the outside world. For some of the residents we interviewed, the focus of their lives remained very firmly centred on family and life outside the home. While the frequent presence of family bridged a gap and helped to relieve the sense of isolation and loss, it could also be unsettling. One resident, for example, whose four daughters visited several times a week, did her laundry and took her to the house of one of them for her weekly bath, confided that her present position might not be permanent. Another, who was visited by one of her seven children every day, also saw the move as a temporary one and talked (somewhat unrealistically) about returning to live in a flat with the support of her daughters. We came across many examples of residents using their family ties to distance themselves from life in the home. An interview with one resident was dominated by references to her family, who also visited daily and took her out regularly. When asked whether she had made friends within the home, she dismissed the need for them:

> No, we just see each other at meals – and they have family as well. ... I've never needed them. I have my family and grandchildren. ... I have my family – it's plenty.
>
> (RI/70)

Like many others, however, this resident acknowledged the role her family had played in her admission to the home and, although she claimed to be happy, elements of resignation, stoicism and self-sacrifice surfaced when she volunteered the fact that 'I am here because of my family. ... It's the only way. They've got their own families' (RI/70). For some who no longer had such close ties with family, the difficulty in adjusting to the reality of their situation could be particularly hard. One man, spoke with enthusiasm about his current daily life in the home – rowing on the river with his wife, walking with the dog and playing with the children – although a photograph of his deceased wife's grave was displayed on the wall together with one of his former dog. Visits from and close contact with family was not always enough, however, to alleviate the sense of loss felt by some, especially those who had recently lost a husband or wife. One resident who had six children, several of whom lived close by and visited frequently, continued to grieve and was unable to overcome his depression.

Visits from fellow residents' families could also be important for those without family of their own.

> Lovely to have a number of children (with Grannies here). One gave me a lovely kiss – *years* since I'd had one like it from a 'tiny'.
>
> (RD/13)

But, in addition to visiting and taking residents out, some relatives had significant social roles within the home which, arguably, helped residents to integrate into home life. In one home, the son-in-law of a resident regularly played the piano or keyboard for sing-songs and dancing, and sometimes for special occasions. Other families were active in running fetes, or helping with gardening projects. Relatives also acted as advocates at times and took action when they were unhappy with the way their relative was being treated or the service they were receiving. The family of one intervened, for example, when they realized that staff were not responding sensitively to their mother's recent and sudden loss of sight. Another swiftly sought the intervention of her daughter when she discovered that a male staff member had been assigned to help her bath.

Friends also featured strongly in some residents' lives. One diarist noted: 'I'm lucky that the home is where it is so that I get many visits from my friends and I can also visit them.' There was not one day in his week's diary when he did not receive or send at least a phone call, and usually friends were seen either in the home or when he drove out in his car. The importance of this aspect of his new life was underlined by this entry:

> 9pm–10pm: A quiet hour before bed. Watched TV and read the novel. Nobody rang or came to see me, not even the carers on duty. Went to bed at 10.45.

(RD/20)

Conclusion

In this chapter we have tried to capture something of the daily lives of residents in the 20 homes we revisited and to highlight the similarities and differences between the two time periods. It is important to emphasize that daily life varied considerably in 1958–9 in different homes. Life in a large ex-PAI, for example, was very different to some of the smaller private and voluntary homes, where residents had access to single bedrooms within which they could pass the time, entertain visitors and pursue personal hobbies. We must be cautious, therefore, about generalization. Nevertheless, as this chapter indicates, it is possible to draw out overall changes as well as the persistence of certain features in the everyday lives of residents over the last 50 years.

Turning first to the changes, there is no doubt that in the 1950s residents were on the whole more physically active than they are today. A major difference, therefore, is that in the past the majority were able to get out and to breach the boundary between the home and the community independently, although the choices for many, once outside the home, were constrained by poverty. In 2005–6, far more were confined to the home. This meant that the majority of residents were reliant on others to provide a bridge between the home and the outside world. In the private homes in particular, when compared to other types of homes, there seemed to be less staff available to support residents in this way, making visits to and from relatives and friends all the more important.

Another, related difference is that in 2005–6 residents were no longer expected to or, by and large, even considered capable of work or serious engagement in ordinary daily living activities. The fact that residents are no longer exploited in this way is of course a good thing. However, work in the past offered some residents the opportunity to engage in meaningful activities, as well as enabling them to continue to exercise lifelong domestic skills. Fifty years later, we found such opportunities in short supply. Despite developments in group living in the intervening years (Hitch and Simpson, 1972; Gupta, 1980; Johnson, 1982; 1993), in the twenty-first century the predominant ethos is one of service coupled with an anxiety about risk and its consequences. Group living arrangements, which go beyond the tokenistic involvement of residents in domestic activities, promote interaction between residents and allow more able residents to assist the less able. Our research revealed that, although the functional ability of residents has declined overall since the 1950s, there was still a mix of abilities among residents in the homes we visited so that such arrangements might be possible. However, the privatization of the care home market and the predominance of the hotel model of care mean that group living no longer seems appropriate. Furthermore, although in a few of the homes we visited, the able and less able residents were integrated, in most, the policy was one of segregation or specialization, reducing opportunities for a group living model of care.

Related to these changes is the increased importance attached to recreational activities. In 2005–6, the expectation by those responsible for inspecting social care services was that homes would have a daily programme of activities and employ a dedicated activities co-ordinator. Examining photographs from the time together with Townsend's descriptions of group activities in the homes in the 1950s, we have detected a change away from activities often initiated by residents, such as card schools, dominoes, singing and piano playing to activities initiated and led by staff. There was also some evidence, in 2005–6, of an uncritical (sometimes infantilizing) and tokenistic approach to activity sessions as revealed by the quality of staff participation and types of activity offered. While some residents enjoyed these organized sessions, others did not and on the whole individual pursuits seemed more popular. These too have changed overtime. Whereas in the 1950s many of the women undertook crochet, embroidery or knitting, in 2005–6 this kind of close work was unusual, mainly because of poor eyesight or dexterity, but also perhaps because fashion in women's pursuits has changed.

Despite these changes, there are continuities. As in the 1950s, it is really the personal relationships – with each other, with staff and with family or friends outside the home – rather than activities per se that sustain residents. However, the two are not unconnected because it is through shared activity that people are able to interact and develop relationships. In 2005–6, there was far more opportunity, particularly in the former local authority homes, for daily life within the home to be conducted in some privacy, sometimes creating a tension for staff as to the extent to which the community side of home life should be fostered and encouraged. Inevitably, there is a potential clash between the drive to de-institutionalize home life but at the same time to create a sustaining community.

8
The Quality of Care

Thus far, we have considered some of the features of continuity and change in the surviving homes and in the everyday lives of those living and working in them. A question that is frequently asked, however, is whether the homes today are better than they were in the past. In this chapter, therefore, we present our findings about the quality of care in the surviving homes then and now. The quality of care in the homes together with policy and practice regarding the registration and inspection of homes was one of Townsend's key concerns. In the late 1950s, homes owned by the local authority were not subject to registration and inspection procedures. Voluntary and private homes, however, were required to be registered with, and inspected by, the local authority. However, Townsend found that for a variety of reasons some of the voluntary and private homes he visited were exempt; others had not been inspected for at least a year and some for more than five years. The quality of inspection was also mixed. Given some of the conditions found by Townsend, this situation was of considerable concern to him.

As outlined in Chapter 2, there has been a sea change in approaches to the regulation and inspection of residential care homes for older people since Townsend's research was published. For example, while we were undertaking our research, it was a requirement that all homes (including local authority homes) be inspected at least twice a year, with one of these visits being unannounced.[1] In addition to the national minimum standards against which the quality of care in a home is judged, staff are now required to have CRB checks (such a requirement also applied to us as researchers). Homes are also required to have a complaints procedure and there are further procedures to protect vulnerable adults from abuse.[2] The main question this chapter addresses is, therefore, what difference do these changes appear to have made to the quality of care and quality of life in the surviving homes nearly 50 years on? As mentioned in Chapter 1, Townsend devised a quality measure to assess the quality of care provided by each home. So we begin by using this measure to compare the quality of care in both 1958–9 and 2005–6 in the 20 homes we visited.

This is followed by a comparison between Townsend's quality ratings and those of the CSCI in 2005–6 in all 37 surviving homes. As one would expect, in many ways Townsend's measure is very different to today's equivalent. It includes, for example, such items as the ratio of wardrobes to beds, of washbasins or wireless sets per head. So comparing the two measures then and now and what they reveal vividly illustrates change and continuity both in the quality of care and in the way that quality of care has been assessed over time. CSCI reports were of course available on all 37 surviving homes but they do not provide the necessary information to implement Townsend's measure, for example, details relating to furnishings or to staff attitudes. It was only possible, therefore, to apply Townsend's measure to the 20 homes we visited.

Townsend's quality measure

In order to make a judgement about the quality of the 173 homes he visited, Townsend devised a 48-item quality measure. The 48 items were organized into five aspects of institutional life: physical amenities, staffing and services, means of occupation, freedom in daily life, social provisions. His book contains a detailed appendix explaining how each of the 48 items was weighted to measure the quality of homes (Townsend, 1962, Appendix 3, pp. 477–91). He gave each home a score out of 100 and placed it into one of six categories ranging from bad at one extreme to very good at the other (see Table 8.1). We used this measure to rate 19 of the 20 homes we visited in 2005–6.[3] The 1959 data are not available for individual homes, only for aggregated categories. Despite this, we were able to draw some conclusions about continuity and change in relation to the quality of care home provision, and about how it is measured.

Given the changes that have taken place since 1959, it is hardly surprising that, as Table 8.1 shows, all but one of the homes in 2005–6 would have been rated by Townsend as good or very good. Although we do not know how the 19 homes were rated in 1959, we can still make some limited inferences by matching our data to Townsend's as shown in this table. All ex-PAIs, for example, were rated as poor, very poor or bad in 1959, which means that the one in our sample which we

Table 8.1 Scores using Townsend's quality measure by 1958–9 tenure, 1958–9* and 2005–6†

1958–9 tenure		0–29 (bad)	30–9 (very poor)	40–9 (poor)	50–9 (fair)	60–9 (good)	70 or over (very good)	Total
		(n)	(n)	(n)	(n)	(n)	(n)	(n)
Ex-PAI	1958–9	8	23	8	0	0	0	39
	2005–6	0	0	0	0	0	1	1
Other local authority	1958–9	0	6	24	20	3	0	53
	2005–6	0	0	0	0	4	1	5
Voluntary	1958–9	0	4	5	13	14	3	39
	2005–6	0	0	0	0	4	3	7
Private	1958–9	1	5	11	15	10	0	42
	2005–6	0	0	0	1	2	3	6
Total	1958–9	9	38	48	48	27	3	173
	2005–6	0	0	0	1	10	8	19

Note: * The 1958–9 data are for all 173 homes visited (Townsend 1962, Table 35, p. 214).
† The 2005–6 data are for 19 of the homes revisited in 2005–6.

revisited must have changed for the better by 1959 standards. Likewise, two of the five former local authority homes and three of the six former private homes show at least some improvement. There is evidence to suggest therefore that, on the basis of Townsend's measure, some of the surviving homes are an improvement on the past and that it may not be just the better homes in 1959 that have survived. We return to this latter issue later in this chapter.

It would be surprising if we had not encountered some difficulties in using Townsend's measure in a modern day context. These difficulties are symptomatic of social change generally as well as changes in care home management and practice. Table 8.2 shows the scores obtained on the five aspects of home life used in Townsend's measure and in what follows, we discuss some of these difficulties and, where appropriate, how these scores compare with the broken down scores in 1958–9.

Physical amenities

This section of the measure includes 11 items relating to the physical environment and it was perhaps the easiest section to apply today. Most of the homes we visited rated highly on these measures, as Table 8.2 shows, with an average score of 20.2 (out of 25). The highest average on this item in 1959 was 15.0 for the voluntary homes and the lowest was 4.8 for the large

Table 8.2   Average scores for 2005–6* and 1958–9† using Townsend's quality measure

	2005–6	1958–9	1958–9 range across tenures	Maximum possible score
Physical amenities	20.2	11.5	15.0–4.8	25.0
Staffing and services	12.3	9.4	10.7–5.8	22.0
Means of occupation	5.2	5.2	6.1–3.2	13.0
Freedom of daily life	11.8	7.4	9.9–2.1	13.0
Social provisions	19.5	14.4	15.3–13.4	27.0

Note: * The 2005–6 data are for the homes revisited in 2005–6.
† The 1958–9 data are for all 173 homes visited (Townsend 1962, Table 36, p. 215).

ex-PAIs (Townsend, 1962, Table 36, p. 215). What these figures reflect, therefore, are the substantial improvements in physical environment and amenities over the years that we have described in earlier chapters such as the move away from shared rooms to single rooms, the exclusive use by individual residents of items of furniture such as bedside tables and wardrobes, the provision of en suite facilities and central heating. Interestingly, the homes scoring below average in 2005–6 are those that had no, or only partial, en suite facilities, such as some of the older purpose-built homes. At the time of their construction they were considered to be of very high quality in terms of physical plant, containing as they did a substantial proportion of single rooms. None, however, had en suite facilities and this is still the case. So what was at the forefront of physical provision 20 and more years ago is now unsuitable and outdated.

Staffing and services

Some of the eight items in this category were more difficult to apply and figures here should be treated with caution. Trying to estimate the ratio of 'staff concerned with direct personal care to total number of residents' was not always easy. One home, for example, employed 85 members of staff, 44 of whom were care staff, some of whom were 'bank' staff which meant they were called upon as and when required. The home also employed agency staff. All these people worked a varying number of hours per week.

Another item in this section relates to visits by doctors: 'routine visiting days per month as ratio of residents unable to go out'. Townsend wanted to measure the frequency of medical attention as related to medical need. In 2005–6 it was unusual for doctors to make routine visits to care homes, although there was one home in our sample where the doctor held a weekly surgery and this was scored accordingly. This meant, however, that most homes scored 0 on this item because medical attention was only supplied on demand and we were unable to make a judgement about how adequate that attention was.

The average score for 'staffing and services' was 12.3 which compares with 1958–9 scores ranging from 5.8 in the large ex-PAIs to 10.7 in the voluntary and private homes (Townsend, ibid.). The apparent improvement no doubt reflects the increased infirmity of residents in 2005–6 and the concomitant need for higher resident/staff ratios.

Means of occupation

This section contains six items relating to how active and occupied residents are and this was relatively straightforward to complete with the exception of the last item 'Occupational therapist' and the proportion of residents receiving regular occupational therapy. In 1958–9 occupational therapists visited or worked in some homes to provide occupational therapy 'classes' which, it would appear from Townsend's book consisted of handicrafts such as basket work or knitting blanket squares. Nowadays, occupational therapists fulfil a different function of assessing and helping people with 'activities of daily living'. There was one home in our sample where an occupational therapist called on some residents once a week. The remainder scored 0 on this item because occupational therapy as it exists in 2005–6 was not provided on a regular basis. However, many homes (as we discussed in earlier chapters) employed an activities' organizer who fulfilled a similar role to that of occupational therapists in 1958–9, albeit a rather more imaginative one, that included exercise classes, chairobics, music and movement, art therapy, reminiscence, quizzes, bingo and dominoes. We decided that the impact of such a person would be reflected in the rating we ascribed to another item in this section on the proportion of residents engaged in individual 'pastimes' and 'group activities'.

The average score (out of 13) for this section was 5.2. This compares with average scores ranging from 3.2 for the medium sized ex-PAIs to 6.1 for the voluntary homes in 1958–9 (Townsend, ibid.). So it would appear that residents are about as occupied today as they were in the past. Townsend commented on the failure of occupational therapy to meet people's occupational needs.

Administrators agree ... that occupational therapy is desirable in residential Homes for the aged and handicapped. Yet such therapy can rarely be organized successfully for more than a tiny proportion of residents, not because the residents are too infirm or apathetic but largely because their interests, talents and capacities are too diverse to conform to a single scheme.

(1962, p. 384)

Data from interviews with managers show that, in 2005–6, although engagement in individual pastimes[4] is more frequent than in group activities, there was nevertheless only one home where less than 20 per cent of the residents took part in some group activity. Our interviews with residents together with our general observations suggest, however, that it still appears to be the case that, despite the employment of activities' organizers who co-ordinated a range of activities in at least half the homes we visited, only a minority of residents' 'occupational interests, talents and capacities' were being met through these organized activities.

Freedom in daily life

This section contained eight items about choices and restrictions affecting daily life such as visiting hours, personal possessions, doctors, bed times and bathing. The average score here out of 13 was 11.8 compared to 1958–9 scores ranging from 2.1 in the medium sized ex-PAIs to 9.9 in the voluntary homes (Townsend, ibid.). As indicated in Chapters 6 and 7, certain freedoms and choices have increased over the years, particularly, as Townsend's low scores show, for those in 1958–9 living in ex-PAIs and other local authority homes. Most of the homes we visited achieved the top scores on the following five items:

- No restrictions on visiting times except meal times.
- New residents being allowed to bring at least one item of furniture.
- Flexible getting up and going to bed times.
- Being allowed to rest on the bed in one's room in the afternoon.
- Being allowed to stay in bed if requested.

This reflects a change in attitude over time, although there may still be institutional practices and routines that prevent the full implementation of these basic freedoms.

What scored less highly in some homes was

- at least ten per cent having a doctor other than the one appointed by the home.

This was usually because there was only one general practice in the area from which the home could secure medical services.

The item that in the vast majority of cases scored 0 was

* being unsupervised while bathing unless unable or unwilling to bath alone.

This was an issue Townsend was particularly concerned about: that people had to suffer the indignity and intrusion of supervision even when it was not required. In our visits in 2005–6, we came across residents who did not require supervision with bathing but nevertheless had to accept it. This is symptomatic of the risk culture that we found dominated many aspects of daily life in the homes we visited. In 2005–6 daily life may not be restricted in some respects as it was in the past but it is restricted in new ways that do not feature in Townsend's measures: not being allowed to go into the kitchen, to make a cup of tea, to go out without an escort, to smoke in your own room, to manage your own medication, to bring your cat or dog to live with you, for example. Had Townsend taken account of such concerns in devising his measure, then residential care may have appeared much less restrictive then when compared with now. What the differences in scores demonstrate is a change in values and attitudes but we cannot conclude that the higher scores in 2005–6 mean that life today is less restricted.

Social provisions

This section contains 15 items including the location of the home in relation to the local community, public transport, holiday arrangements, birthday celebrations and entertainments, religious services, availability of radios and TVs, managing death and dying. The average score for this section was 19.5 compared to scores ranging from 13.4 in the medium-sized and small ex-PAIs to 15.3 in the voluntary homes (Townsend, ibid.).

The first item that posed some difficulty for us was 'Provision for married couples'. We were not certain what the word 'provision' was alluding to and Townsend does not find it necessary to explain. In 1958–9, many of the homes were either single-sex or men and women were segregated. Provision could simply mean allowing married couples to live in the same home. However, we took this item to mean a double room: one that could be shared by a married couple. There were homes that we visited which accommodated married couples, but only in single rooms.

Two further items signalling change were about the ratio of radios and television sets to beds. In 1958–9 the possession of a radio was much more common than a television, the latter being extremely expensive.[5] Hence, whereas Townsend gave the top score to a home with a ratio of 1:4.9 or less radios to beds, the equivalent score for televisions to beds was for a ratio of 1:29 or less. In 2005–6, most rooms in most homes included a television set either belonging to the resident or to the home, and all had at least one communal television set and associated recorders and players. Not surprisingly, the availability of telephones was not an item on Townsend's schedule, neither was the availability of personal computers. Although, not used in this quality measure, we noted some significant variations in access to personal telephones. In one voluntary home catering predominantly for the affluent middle class, all residents had their own phone whereas in one of the local authority homes, only one resident out of 27 had a telephone.

A further item signalling change related to holidays. Underlying this item in Townsend's schedule was his assumption that a proportion of residents in every home was likely to take holidays. So homes in which the proportion was under 40 per cent scored 0. In 2005–6, the number of residents taking holidays away from the homes was miniscule: residents are no longer expected, or in the main considered able, to go away on holiday. Consequently all the homes we visited scored 0 on this measure.

Other items in this section have worn the test of time, for example, those relating to public transport and the siting of the home in relation to the nearest urban centre. Although there may have been more residents able to use public transport in 1958–9, it remains an issue for some residents and for visitors and staff, despite the increase in car ownership during the intervening years.

Summary

We can infer from Table 8.2 that, by 1959 standards, there have been improvements in the standards of care provided in the surviving homes in all but 'means of occupation'. This is particularly the case for physical amenities and for social provisions where the difference between the 2005–6 average scores and the maximum averages in 1959, and the maximum possible score, has changed the most. Does this represent evidence of a substantial improvement in residential care over the last 50 years? We remain sceptical. Although there have undoubtedly been dramatic improvements to the physical environment in most homes, as discussed above, many of the differences we have recorded tell us as much, if not more, about social and cultural change in the way quality is represented. An examination of findings from the use of a modern quality measure amplifies this point.

Photograph 8.1   Dayroom 1959 and lounge 2006
Source: The Peter Townsend Collection; *The Last Refuge* revisited collection. © Jean Corston
(top), © Julia Johnson (bottom)

The CSCI quality ratings

The decision to use the CSCI inspection reports to reach a modern quality score for each home meant that we could apply this measure to all the surviving homes and not just those we visited. However, of the 37 surviving homes, four were located in Wales, three of which we visited. Following devolution, the care standards used by care home inspectors in Wales changed. We therefore excluded them from our analysis and rated only the 33 surviving homes in England (including the 17 we visited).

The CSCI standards

The relevant CSCI inspection reports assessed each home against 37 of the 38 standards (see Appendix 4). The least frequently assessed standards were standards 11 (dignity in death), 17 (protection of legal rights) and 34 (suitable accounting and financial procedures). We sometimes had to go back two years before finding an assessment of these standards. In one instance, standard 34 had not been assessed for over three years despite the fact that, when it had last been assessed, it was given a score of only 1, the lowest it could achieve.

There were various factors which were likely to affect how homes were assessed against these standards. First, although they are national standards, there was still a possibility of local variation in their application by inspectors from different CSCI offices. Secondly, the judgement of individual inspectors was likely to vary. The ratings of one of the convents, for example, fell significantly following a change of inspector. Conversely, the ratings of another home improved when a new inspector was appointed who, according to the manager, developed a better working relationship with him. Likewise the manager of one of the faith-based homes was pleased to have a new inspector who understood better the religious ethos of the home; a previous inspector had wanted them to introduce bingo which as far as this home was concerned was regarded as gambling. Thirdly, we noticed that homes were affected by changes in ownership and management. One of the private homes, for example, went through a bad patch in 2004 and 2005 and several standards were not met. By 2006, under new management, a vast improvement was recorded. Our measure may not have reflected some of the changes therefore, either good or bad, that had taken place between the date when a particular standard was last assessed and the date we gave the home an overall rating.

In the inspection reports, a home was given a score by the CSCI inspector ranging from 1 to 4 for each of the 37 standards. By totalling

Table 8.3   Percentage of CSCI standards exceeded, met and not met, England, 2005–6

| | 2005–6 tenure | | |
	Voluntary homes	Private homes	All homes
Standard exceeded (commendable)	7.2	0.9	4.2
Standard met (no shortfalls)	80.4	69.6	75.4
Standard almost met (minor shortfalls)	10.7	25.1	17.4
Standard not met (major shortfalls)	1.8	4.4	3.0
Total (= 100%)	627	550	1214

these scores for each home, we were able to allocate an overall score out of a possible maximum of 148.[6] As well as being able to compare how quality was measured by Townsend and CSCI, we were able to use the CSCI scores to analyse differences in quality among the surviving homes in 2005–6. Of the 33 surviving homes in England, ten had been owned by the local authority in 1958–9. By 2005–6, three of these had been transferred to the voluntary sector and six to the private sector. Only one remained in the hands of the local authority. In looking at the differences between the homes in 2005–6, our analysis focused on a comparison between the 15 private and 17 voluntary sector homes.

As Table 8.3 shows, the majority of CSCI scores for homes in both the voluntary and private sector were 3 (standard fully met). This probably reflects the inspectors' expectations and is the benchmark for judging whether a home meets, exceeds or falls below the minimum standard. The table also shows some substantial differences between the two sectors: 12.5 per cent of the scores for the voluntary homes did not fully meet the standards compared to 29.5 per cent for the private sector. The differences are even more pronounced at the extremes: 7.2 per cent of the scores for the voluntary homes exceeding the standard compared to 0.9 per cent for the private sector. Conversely, 4.4 per cent of the scores for the private homes indicate major shortfalls in meeting the standard, compared to 1.8 per cent for the voluntary homes.

Townsend rated the voluntary homes much more highly than the private homes (1962, Table 35, p. 214). Nearly twice as many of the voluntary homes, compared with the private homes, were rated as good or very good. While over a third of the private homes were rated as bad, very poor or poor, less than a quarter of the voluntary homes were. In addition, none of the voluntary homes were rated as 'bad' and none of the private homes were rated as 'very good'. Our findings were very similar. However, the correlation between Townsend's score as applied

Table 8.4   CSCI scores for follow-up study homes, England, 2005–6

	1958–9 tenure	2005–6 tenure	No. of standards scoring				Total score
			4	3	2	1	
St Michael's*	Priv	Priv	0	11	21	4	79
Fairfax House	LA	Priv	0	17	18	2	89
Laburnum House	Priv	Priv	0	21	13	3	92
Elm Lodge*	LA	Priv	0	25	7	4	93
Beaconsfield	Priv	Priv	0	21	14	2	93
Hill House	Priv	Priv	0	20	17	0	94
St Augustine's	Vol	Vol	0	26	5	6	94
Monmouth House	Priv	Priv	0	24	11	2	96
Loxley NH*	Priv	Priv	1	26	6	3	97
Longacres†	LA	Priv	0	28	7	0	98
Edenfield Lodge	Vol	Vol	0	27	7	3	98
Carlton House	Vol	Priv	0	27	9	1	100
Pine Grange	Priv	Priv	0	30	4	3	101
Greenacres	LA	Priv	0	29	7	1	102
The Cedars*	LA	Vol	0	30	6	0	102
Overton Court	Vol	Vol	1	27	8	1	102
Bridge House	LA	LA	0	29	8	0	103
Mulberry House	Priv	Priv	0	31	6	0	105
Hillcrest	LA	Vol	1	29	7	0	105
The Brambles	Vol	Vol	1	29	7	0	105
Acton Lodge	Vol	Vol	0	31	6	0	105
St Mary's	LA	Vol	0	33	4	0	107
Richmond Lodge*	Priv	Vol	1	33	2	0	107
Avebury Court	Vol	Vol	0	34	3	0	108
Newholme	Vol	Vol	4	27	6	0	109
The Gables	LA	Priv	0	36	1	0	110
The Limes	Vol	Vol	1	35	1	0	111
Springfields	Priv	Vol	2	35	0	0	113
Manor House	Priv	Priv	4	33	0	0	115
St Peter's	Vol	Vol	8	27	2	0	117
Rivermead	LA	Vol	7	30	0	0	118
Oakfield Court	Vol	Vol	8	29	0	0	119
The Laurels	Vol	Vol	11	26	0	0	122

Note: The total possible score for 37 standards is 148.
* Five homes with scores based on only 36 standards.
† One home with scores based on only 35 standards.

in 2005–6 and the CSCI score for the same 16 homes is 0.10, indicating that it is positive but low. This suggests that the two scores are to a considerable extent measuring different aspects of quality. So we cannot assume that the differences between private sector and voluntary sector provision have simply persisted since the late 1950s.

Table 8.4 shows the total scores we gave each of the 33 homes and below we discuss some of the worst and the best, using the archived reports, our own and CSCI's to compare them over time.[7]

The worst homes

The home with the worst CSCI score was St Michael's, a small private home accommodating 15 residents and situated in a large urban conurbation.

The archived report on this home indicates that it opened in 1958 and was visited by a researcher in 1959 when it was reported to house eight women residents, seven of whom were over pensionable age. There was one four-bedded room, a three-bedded room and a single bedroom. The home had no central heating. During the visit, five of the residents were 'sitting up in bed': three had been in bed for seven or eight months and another for two months. The matron worked from 6.30 a.m. until midnight with the assistance of two domestic staff and her husband (when he was not working at his job). One resident said she wanted to leave because of matron's very bad temper, adding that 'nurse carries on something dreadful'. Although birthdays were celebrated, there were no group activities or outings and none of the residents had any hobbies. Only one resident ever went out and apart from visiting relatives, there were no callers to the home: neither chiropodist, hairdresser, occupational therapist, social worker nor clergyman.

Despite several approaches from us, we were unable to obtain consent from the proprietors to visit this home. They had owned it since 1985. In 2005, they turned their business into a private limited company and later that year acquired a second home. For some time, one of the proprietors was managing both homes. Although a new manager was appointed to St Michael's towards the end of 2005, the manager post for the second home was still vacant in April 2007. Both homes were inspected by the same inspector and both were poorly rated. St Michael's met less than a third of the standards and, at its last inspection, 26 statutory requirements were laid down in order to meet the care standards. By way of example, four of the five health and personal care standards had not been fully met. An enforcement order had already been issued by an inspecting pharmacist regarding the administration of medicines. Staff were reported to be keeping inadequate records on a resident whose daily intake of food needed monitoring. They were also criticized by the inspector for failing to observe adequately the privacy and dignity of residents by using inappropriate descriptors of residents in the daily record book such as 'snappy', 'sarcastic' and 'in a right paddy'. One member of staff was observed to speak abruptly to a resident during the inspector's visit. Three of the four staffing standards were unmet and concern was expressed about low levels of training and qualifications and inadequate checks on new recruits.

While our information on the quality of care provided by the home in 2005–6 was limited to the CSCI reports, it would appear that, as in 1959, this home still offered a poor quality service. The other home that failed to fully meet the majority of the standards (fully meeting only 17 out of 37) was Fairfax House, another private home located in the south-west of England belonging to one of the largest corporate owners in the UK and specializing in dementia care. In 1958–9, it was a converted Georgian mansion belonging to the local authority. It had been acquired by the council in 1948. When visited by the researcher in 1959, it accommodated 78 residents, some in prefabs (originally erected as temporary buildings) in the grounds. The researcher's report at the time noted:

> All in all, my general impression of [...] was that this was a disastrous mistake on the part of the Welfare Committee some ten years ago in buying the place. It has saddled them with innumerable difficulties ever since, although it seemed to offer the means of a cheap site with suitable ground-floor accommodation at the time. It is much too far from the city, and has consequent difficulties with staff. [...] It should be noticed that while this accommodation is not included under our former Public Assistance Institutions, it is as bad, if not worse than many of those.
>
> (PTC, Box 36, HI 4)

In 2006, Fairfax House accommodated 48 residents. It was acquired by the company in 2002 (although it subsequently merged with another company) when, following its first inspection under the Care Standards Act, it failed to fully meet 22 of the 38 standards. Only one of the eight standards relating to the environment was met. Nearly four years later, when our own research was conducted, it still failed to fully meet 20 of the minimum standards with only two of those relating to the environment being met. A series of inspection reports referred to peeling paintwork, torn wallpaper, cracked window panes, windows that wouldn't open properly, fire doors that wouldn't close, worn carpets, exposed wiring and unsafe infection control practices. In considering these reports, it is important to remember that, being run by the local authority, Fairfax House would not have been subject to inspection and regulation prior to 1992.

In June 2004, the inspection report concerning staffing (where three out of the four standards were not fully met) and management and administration (where six of the eight standards were not fully met) noted that the home was staffed with the minimum requirement and that it needed to ensure that there was adequate supervision of vulnerable residents at night. In particular it noted that staff 'are not turning up for work, particularly at the weekends and this is having a knock on effect'. The new manager, it noted, was working on matters to do with training, supervision and staff support. The report went on:

> At the time of this inspection the new Manager has been in post for 4 weeks. She has inherited many issues, which have been the end result of several Manager changes within the Home and changes at company regional level. The company's auditing and quality assurance has been brought into question in this report. All financial transactions are recorded and processed satisfactorily by the Home's Administrator. The record keeping and filing generally in the Home has needed to be organized and in some areas has been non-existent. This is in the process of being rectified. The Inspector was not able to evidence safe working practices in all areas.

Eighteen months later, in January 2006, concerns about staffing levels at Fairfax House continued to be voiced.

> The home has set its own agreed staffing limits, however from reading the staff duty rotas the home has been running under their agreed limits. The majority of comments received from visitors about the home are that they felt there is not enough visible staff to meet the needs of the service users.

In the same report, continuing concerns about the laundry were raised together with other environmental issues.

> There were numerous maintenance issues relating to the laundry. Due to the seriousness of the situation this was reported the Health and Safety Executive. Protective clothing was not available in the laundry area. Concerns received from relatives of the service users is that service users do not have many of their own clothes in their wardrobes and are often wearing other service users clothes. This

issue has been ongoing for long period of time and the arrangements of the laundry need to be reviewed to address these concerns.

The report also noted residents wearing other residents' clothes, a 'large number' of missing chair cushions and that smells from some rooms were apparent on entering the home. By May 2006, a number of issues were raised in the summary of what Fairfax House 'could do better'. In addition to the laundry service and the presentation of meals, these included the review of care plans, mentioning that the plan for one 'service user' had not been reviewed since November 2004. Another issue concerned the lack of planned activities and the report noted that the home was trying to recruit an activities co-ordinator. It also reported that since the last inspection of Fairfax House, the standard of vetting and recruitment practices had declined 'potentially leaving service users at risk'. This worrying report should also be set in the context of the home's fees which were reported to be between £460.75 and £710.00 per week.

Both St Michael's and Fairfax House were private for-profit businesses, but very different homes in terms of size, location and the type of company that owned them. What they shared in common, however, was that they were poorly perceived by the researchers who visited them in the late 1950s and by the CSCI inspectors who visited them several decades later, leading us to tentatively conclude that the worst homes in 2005–6 had been among the worst in 1958–9. This demonstrates how difficult it is to change a culture of care that is based on a long-standing history of poor practice.

The best homes

Of the 33 homes included in Table 8.4, five fully met or exceeded all the standards. Four were voluntary sector homes, one of which had been an ex-PAI in the late 1950s; the fifth was a private sector home.

The Laurels was the highest scoring home of all with 26 standards fully met and the remaining 11 exceeded. It was located in the south-east of England and accommodated 33 residents including people diagnosed with dementia. It was one of 16 homes in the south, midlands and East Anglia owned by a charitable organization. In 1959, it belonged to the Women's Voluntary Service (WVS). The Laurels is a large Victorian house converted for use as a home in 1953. Robert Pinker, who visited it in 1959, reported that all the bedrooms were single (except for one double room), had their own wash hand basins and were on the ground floor. He described 'the standard of provision' in these rooms as 'very high indeed' with 'an abundance of personal effects on show'. He also described a relaxed regime with a variety of activities in which individual residents were involved and a residents' committee set up and run by the residents. He was 'very impressed by the calibre and personality of the Matron' and her 'attitude towards personal possessions and the fact that all residents have their own room' as well as 'their own committee' (PTC, Box 36, HI 11).

Overall, Pinker's report is extremely positive and there are no overt criticisms. Interestingly, however, the report makes reference to some attitudes and practices which in 2005–6 would not be considered very acceptable. For example, on the subject of the kind of people admitted to The Laurels, the matron is quoted as saying 'We try to keep the same sort of people – decent clean people who can hold an intelligent conversation.' Clearly the home did not cater for those requiring much personal care either: 'We can't keep the incontinents. ... We don't do actual nursing. If they needed that they would go to hospital or a nursing home. We would keep a senile case if she wasn't too bad.' The fact that Pinker should view the matron and her attitudes so positively is evidence of the extent to which values have changed over time.

We did not visit The Laurels so we do not know whether such attitudes and judgements have persisted. However, the report of the unannounced CSCI inspection on this home, conducted in November 2006, was also extremely positive. Its summary is comprehensive covering the home's information pack, the reviewing of residents' care plans, access to health care, social activities, choice of meals, complaints, the personalization of rooms and the training of staff. It notes that the home 'will not admit any prospective residents until they are satisfied that the individual's needs can be met'. The following details on 'Daily Life' provided in this inspection report, quoted at some length, offer interesting evidence of what is highly valued in 2006 in terms of quality of life for residents.

> The home offers a wide range of activities for residents to choose from which include flower arranging, informative talks, bingo, hoopla, dice, knitting, ludo, snakes & ladders, guitarist, coffee mornings, painting, knitting, quizzes, church services. Residents are encouraged to participate in communal activities to enable them to develop new friendships and build up camaraderie. On the afternoon of the inspection visit several residents, were observed to be having a hand massage.
>
> In discussion with the deputy manager they reported that residents had the opportunity to celebrate Halloween as the home had a party

and that Father Christmas visits the home on Christmas day. There was photographic evidence of outside entertainers performing in the home which included dancers and singers and a visit from a donkey which belongs to an ex-member of staff. The home publishes a quarterly newsletter which gives details of what activities have been provided. The home is actively trying to raise money for a mini-bus.

Many readers of this might feel uncomfortable with some of the charitable, and perhaps infantilizing, aspects of this summary: snakes and ladders, donkeys, Father Christmas, for example. The report of daily life continues, however, with a series of very positive observations about the way the home facilitates choice regarding movements in and out of the home, contact with friends and family, meals and other aspects of daily life. For example, it is commented that 'Wall safes are available in all bedrooms for the safekeeping of resident's individual monies for them to manage.' The way the inspector evidences her observations is interesting.

> Lunch is served in the dining room, however those residents who prefer can take their meals in their rooms. Those residents requiring assistance with eating were observed to be seated in a quiet annexe area just off the lounge to afford them privacy and dignity. The home offers a set menu at both lunch and supper time (tea) but alternatives are available if residents do not like what is on the menu. A member of staff was observed informing residents what was for lunch and asking them what they would like, in talking with the senior member of staff they confirmed that this is done on a daily basis.

There is little doubt that this home, when compared to St Michael's and Fairfax House, was offering a superior and safer service. What is more, this service, which was being provided in a similar if not more affluent and desirable location than Fairfax House, was being offered at almost half the price with fees, in 2006, ranging from £355.11 to £465.00 (the maximum charge being only £5.00 less than Fairfax House's minimum charge).

The Laurels is an example of a home that has continued to provide what is judged as an exemplary service. But what about the other homes which likewise fully met or exceeded 100 per cent of the standards? Do they likewise match the hypothesis that what is rated highly today was rated highly in the past?

Oakfield Court, the next highest scoring home, certainly supports this hypothesis. This was and still is owned by a Quaker organization and the researcher who visited it in 1959 describes 'one of the old ladies trimming the edge of the lawn'. He also describes the handicrafts group and a choir 'to which half the home belongs and which meets weekly under the conductorship of a local Quaker'. In summing up, he comments:

> Why is this the best home I have seen? Three reasons: (1) The high standard of material provision (2) the organisation in grouped dwellings and (3) the character of the Warden who seemed to me to be ideally suited for his job. [...] The sort of incident which influenced my evaluation ... was this: as we were sitting finishing our lunch, one old lady on her way out stopped by the warden's chair for a few minutes. They chatted and talked and laughed together for a while without any apparent condescension on one side or embarrassment on the other. In how many homes would this happen, I wonder? In how many homes anyway do the staff eat with the residents?
>
> (PTC, Box 36, HI 7)

The October 2005 inspection report noted that Oakfield Court was 'an established, well managed and well maintained service that continues to provide high quality care and accommodation for older people'. It also noted thorough staff recruitment procedures which 'ensured the protection of service users' and that staff received effective training and regular supervision.

Manor House was the only private home to fully meet all (and in some instances exceed) the minimum standards. This is another interesting example of continuity. In 1959 it was reported that this home had been run as a private hotel since 1938. In 1955, the owner was obliged to convert it into a residential care home because the majority of his residents had reached an age where they required some personal care. He clearly found Townsend's visit tiresome and was reluctant to identify with residential care.

> 'This isn't an institution, you know – not one of those council hospitals, or whatever they call them'. Another one of Mr C's remarks was 'We take your social class ones – that's why we're all single rooms here. Your working class people like to be huddled together.' 'We don't take your threes, fours and fives, and we don't want them either.'
>
> (PTC, Box 36, HI 11)

Townsend's report notes that 'this is the most luxuriously appointed home I have visited in this survey'. Despite his obvious reservations about the owner, he also notes that staff knocked on doors and waited for a response before entering. His concluding paragraph sums up Manor House as follows:

At fees between 10 and 30 gns [guineas] a week there is a waiting list for admission to this Home. Of course the standard of amenities and service – especially nursing care – could not really be bettered, and an old person entering this Home, and able to pay the fees, can remain there even as her health progressively deteriorates. You could hardly call the proprietor a sensitive man – except when it comes to making money, but then he certainly gives value for the fees he charges. I suppose this Home illustrates the truism that if you've got money you can buy anything and even acquire the means of dying with a relative degree of dignity. If you haven't got the money you take Hobson's choice and make the best of it.

(PTC, Box 36, HI 11)

In February 2005, according to CSCI inspection report, money still bought residents of Manor House 'Dignity in death' (standard 11, fully met). By this time it belonged to a family-run company which owned several other homes in a southern county of England. According to its brochure, Manor House offered

an exceptionally attractive residence for retired ladies and gentlemen ... [It] is designed to care for the needs of those who wish to be in pleasant surroundings, with the assurance of high quality cuisine and the attendance of helpful and caring staff during the day and night. [...] The regency dining room has a personal table service. Meals are served in the dining room for all residents unless room service is preferred. Family and friends are also welcome to stay for luncheon or afternoon tea in the private dining room.

(Brochure, November 2006)

It is clear, as it was in the past, that the hotel model of care is emphasized and what kind of clientele Manor House was aimed at.

Rivermead and Springfields, the other two homes which met or exceeded all the standards, had both changed tenure since 1959. Rivermead was an ex-PAI and Springfields a small private home in 1959. Both have subsequently been taken over by voluntary sector organizations and both have clearly improved in terms of quality over the intervening years, neither being regarded as good homes in the late 1950s.

Rivermead was located in a rural county town. It was demolished and replaced, in 1963, with a purpose-built home on the same site and residents were moved from the old home to the new one. In 1995, after the NHS and Community Care Act 1990 had been implemented, along with all the other council homes in the county, ownership was transferred to a newly created industrial and provident society run by a voluntary board of trustees. This organization now employs 650 staff in the county and operates 12 care facilities, including the nine (previously county council) care homes. In 2006, Rivermead was registered to accommodate 40 residents including older people with dementia or learning difficulties. In 1959, this was a 'joint unit' with a hospital in the same grounds and it remains so today. We know that none of the ex-PAIs were given a rating higher than 'poor'. So we can only assume it has improved. There is no evidence to suggest, however, that its transfer to the voluntary sector is a key factor in this. Another surviving home (Hillcrest) likewise was an ex-PAI which was replaced by a purpose-built home and then, in 1999, became part of a bulk transfer to another voluntary care trust. This trust was set up in 1991 to take over the 16 care homes in another county. By 2007 it had become the second largest not-for-profit care provider in the UK, operating 74 homes in four counties, responsible for the care of some 3300 residents and employing around 3500 staff. Unlike Rivermead, Hillcrest was not one of the top scorers, suggesting that the 'voluntary' status of Rivermead is not a factor in its success. However, whereas the trust owning Rivermead is locally based with an identity associated with the county within which it is located (indeed it could be called a quasi-local authority), the trust owning Hillcrest is not dissimilar to a large national for-profit company.

Springfields was taken over in 1973 by a faith-based organization which at the time of our research owned eight other care homes across the south of England and East Anglia. The researcher in 1959 was 'not impressed' with the home which accommodated 22 residents. In the main, this was because of the matron's attitude. One resident had been locked into her room and the matron denied the researcher access to potential interviewees. Visiting hours were very restricted and residents were not allowed to entertain their visitors in private. Although the home was described as being in a good state of decoration and repair in the late 1950s, we were told that by the time it was taken over in 1973, the building was almost derelict. The organization that runs it has clearly spent a great deal on extensions and improvements, including the addition of independent flatlets, and for those who accept its doctrinal basis of faith it appears to provide a high quality of care.

Conclusion

By Townsend's 1958–9 measure, the difference in the standard of care then and now in our sample of homes suggests substantial improvements have occurred over time, particularly in the physical facilities and social provisions available to residents. However, it is clear both from our analysis of Townsend's measure and from the contrasts in quality that are revealed by the CSCI ratings, that standards and expectations have changed significantly over the years, and that by contemporary measures the contrasts and inequalities Townsend found in 1958–9 have persisted. In 2005–6 there is still a marked contrast between good homes and bad homes. Our findings also show that money does not necessarily buy the best care. Local authority supported residents are now a much higher proportion of those living in private and voluntary homes, particularly the former. It is possible that the private sector is becoming increasingly polarized between those homes that are insufficiently funded through local authority contracts and those that rely in the main on the personal financial resources of a wealthy clientele.

Our data also suggest that overall the quality of care provided in the voluntary sector homes is higher than that in the private sector, and that the best homes by and large are to be found in the voluntary sector, particularly those that originated in the voluntary sector, whereas the worst homes are to be found in the private sector. Neither sector is uniform, however, and clearly there are important differences in the resourcing and management of homes within both which may account for differences in quality as much as tenure. Furthermore, although the minimum standards used by CSCI are intended to provide an objective and standard measure, it remains the case that assessing quality is essentially a subjective exercise based on individual perspectives, experience and opinion.

The market in residential care has clearly been more volatile in recent decades. It may have been hoped that policies promoting the development of the 'independent' sectors at the expense of local authorities would lead to poor homes closing, surviving homes improving and good homes setting the standard. We were rather surprised that the two worst homes, particularly St Michael's, were still registered given their shortcomings. We noted, however, that the inspection reports for St Michael's categorized all the standards which achieved a rating of 2 (standard almost met) as 'adequate'. This designation suggests that the minimum standard for what is considered acceptable may have been lowered from 3 (fully met) and that 'almost met' means 'good enough'. It would appear, therefore, that the market is not a wholly reliable mechanism for raising care standards.

Part III Conclusions

9
Revisiting and Reuse

In Part II of this book, we have described our findings both from the tracing study and the follow-up study. Our particular design is unusual and as such has both strengths and limitations regarding what we can conclude about policy and practice in relation to residential care for older people. As pointed out in Chapter 1, ours was not a 'restudy' such as that conducted by Stacey et al. (1975), Phillipson et al. (2001) and Charles et al. (2008). Rather it was a 'revisiting' study: one that has revisited not only how Townsend conducted his research, published and archived his findings but also a very specific set of surviving homes. Through access to his archived data, we have been able to visit these homes and compare his findings with ours.

This is the first of two concluding chapters to this book. In these chapters we want to reflect upon what we have learnt from revisiting *The Last Refuge*. Our research was both historical and methodological and the focus of this chapter is on some of the methodological issues it has raised. In it we draw some conclusions about the strengths and weaknesses of revisiting studies and of our design in particular. In the next and final chapter, we review what we have concluded about continuity and change in residential care for older people.

In Part I of this book we identified a number of aims, two of which relate to our methodology. The first concerns our use of volunteer researchers and what this has contributed to our understanding of involving older people in social and historical research. The second concerns the use of archived data and what our experience can contribute to policy and academic debates in this field. Both these aspects of our methodological approach have been the subject of much discussion in recent years and there is a growing literature surrounding each.

Working with 'volunteer' researchers

In the field of gerontology, it is now a common requirement of grant-awarding bodies that older people should be involved in discussions on the research strategy (Ray, 2007). For example, the Growing Older and the New Dynamics of Ageing research programmes funded by the five UK research councils have given such involvement a high priority and Walker (2007) identifies three reasons for this. First the rise in individualism, he argues, has led to the consumer perspective being sought in order to make services for older people more 'responsive'. Secondly, the UK research councils, reflecting the concerns of government, have become increasingly concerned with 'user' engagement and the role of 'non-academic' users of research. Thirdly, with the growth of social movements representing a variety of groups, such as the disability rights movement, there is the demand not just for their involvement but for their control of the research agenda. The third of these reasons has spawned a burgeoning literature on participative research, much of which focuses on the users of health and social care services (Beresford, 2007a, 2007b; Nolan et al., 2007; Miller et al., 2006; Turner and Beresford, 2005; Lowes and Hulatt, 2005). Overall, the goal of user involvement appears to be one of 'empowerment'. However, although older people are major users of care services, this is not the only reason for their involvement in participative research. Barnes and Taylor (2007) suggest four broad types of involvement, one of which is involving older people as 'research practitioners' either working alongside a research team as data collectors, or involved in the research design and analysis and interpretation of the data.

The tracing study involved older people as 'research practitioners', working alongside ourselves, the research team. They were not involved in the design of the research or in the aggregated data analysis. They were, in Barnes and Taylor's terms, 'data collectors'. In other words, they worked to our agenda and the balance of power remained with us, the professional academic researchers. Some might be critical of this form of involvement. Clough et al. (2006), for example, might argue that our failure to involve our volunteers throughout the research process could be detrimental, particularly if we failed to involve them in the dissemination process leaving them disappointed about the outcome of their participation (Ray, 2007). Others might criticize us for treating them as 'intellectual hod carriers' (Drake, 2005, p. 116) or for simply exploiting their willingness to undertake unpaid work.

The gains of volunteering

Information we collected about our volunteers and what they gained from their participation in our project did not support these potential criticisms. As described in Chapter 3, most of our volunteers were well-educated middle-class women who were committed to lifelong learning. Through the personal profile forms they returned to us, they identified three main gains from their involvement: acquiring new knowledge and skills, contributing to something that they felt to be worthwhile, and for some, there were specific gains of personal significance.

The acquisition of new knowledge and skills was the most commonly cited gain and included references to being introduced to new sources of information through local records, developing skills in conducting Internet searches, learning how to approach archival research, learning about the importance of record keeping, learning about the history of residential care, and to increasing their own local knowledge. What came across from the majority was the gain in confidence that the acquisition of new knowledge and skills promoted. As one person put it, it was 'great for boosting self-confidence'. Another commented on how she had become a source of admiration in her local University of the Third Age group and had been asked to write a piece for its newsletter about her research for us. One who had never used a public record office before and for whom the business of trying to locate relevant archives was at first frustrating, wrote a report for one edition of our newsletter. She ended very positively: 'The excitement with which I greeted my partner when that elusive but vital bit of information completed the jigsaw made it all worthwhile for me! Yes, it was worth volunteering.'

But it was not just those who were new to archival research who claimed to have learnt from their experiences. A retired university lecturer, for example, said 'I was volunteered by my local reference librarian, who saw me as a suitable person to do this! However, I was perfectly willing to participate and I thoroughly enjoyed my exploration of some new territory. I have learned quite a lot!' In a piece she wrote for our newsletter, she described the importance of not just the records in the local studies libraries but also of the local library users who happened to overhear her enquiries of the librarian. As she said 'there is nothing local historians enjoy more than imparting information and their ears prick up when they hear the librarian confessing to ignorance'. Consequently, she found several useful informants, such as the person who said 'Excuse me! I know where that is, my mother used to work there. I'll ring her.' Her explanation of what she learnt from the project is worth quoting at some length:

> What all my informants have taught me is that they use their own history to answer questions about the place and time of 'lost' buildings. Events within their own family are remembered clearly and provide time-markers as they visualize the building as it was. Further questions can prompt detailed description and anecdote about long-dead residents. I believe that the researcher needs to have done some groundwork first, having visited the site if possible. [...] A variation on the theme of oral history can then be used to tap into the memories of people who knew Townsend's homes and their inhabitants well over thirty years ago. The chain of memory is a long one because informants refer to an older generation, their parents, for confirmation. They recall, too, their younger selves listening to the reminiscences of former inmates of the workhouse. Two of my informants remember old people who had lived in the workhouse and made the transition when it became a Home that was surveyed by Townsend.

She went on to conclude:

> My research is therefore a joint effort owing much to fortunate encounters with strangers, who could link the old people's homes into their own history. An unexpected bonus for me is the insight I have gained into how purposeful reminiscence can help sort out a sequence of events in a building's history.
>
> (Kemp, 2005, p. 13)

This and other responses demonstrate that, despite their varied experience and expertise, the tracing study provided people with a range of opportunities to gain new insights, knowledge and skills. The responses of our volunteers also offered us new insights so that effectively we were working in partnership.

The second gain was reflected in such comments as it was 'well worth doing' or it was 'good to be using [one's] brain in [a] valuable research project'. Although two volunteers complained about how unhelpful local officials had been and the difficulties in accessing records, a further source of satisfaction which several commented upon was how helpful people had been, not just local studies librarians but also less formal informants such as shopkeepers, vicars and funeral directors. In fact another contributor to one of our newsletters described how some sixth-formers had taken her on a conducted tour of one of her homes which is now a private school. A few commented on the people they had met through doing this research and one volunteer, who was resident in a sheltered housing scheme, said 'Well I've enjoyed it and a number of fellow residents have also been interested and from time to time have asked me how I'm getting on.'

For a few, involvement in our research had a more personal resonance. One volunteer, a former social services area manager, described how it rekindled an old interest. For another, her investigations 'radically changed' her view of residential care for the better. One wrote 'I became dogged in my research, not giving up until I had considered all avenues were investigated. I realized a side of my character otherwise unknown so thank you for the opportunity, If I was younger, I would have taken research as my occupation.' Echoing the theme of self-discovery, another volunteer who agreed to be interviewed for an Open University audio programme, explained how she found that contrary to her former view of herself she *was* able to approach strangers for information with some confidence. A volunteer who contributed to one of our newsletters wrote an account of being introduced to a 105-year-old woman who was able to tell her where the home she was investigating had been and something of its history. Visiting this woman at her house prompted her to reflect on the future for her own parents and on the reality of community-based care for vulnerable older people:

> It was a learning experience for me. She was pleased to be in her own home – as I imagined my own parents would be – and it was good to see a support network working to keep her there. But I became increasingly aware of how vulnerable she was. Various agencies, a stranger, were in and out of her house all morning, she was paying them with cash and she was worrying about them. It made me think.
>
> (Pendleton, 2008, p. 9)

Finally, on a rather different note one wrote:

> Unfortunately my husband became very ill last year and sadly died. ... During his illness and in the immediate aftermath my life was put on hold, but when I noticed that this project was still open towards the end of the year I decided to put my name forward. I have very much enjoyed delving into the records to try to piece together some of the background of these two residential homes and it has been a rewarding exercise. At times the discovery of a new piece of

information has proved quite thrilling. On a personal level it has been therapeutic for me to find something new and absorbing to do at this difficult time.

These responses suggest that involvement in our research was not a disempowering experience for our volunteer researchers. Indeed, the notion that we should be using our research as a vehicle to empower older people was put into question when we attempted to use the final edition of our newsletter to address the issue of closure for those who had been involved in the project. For example, a married couple, who investigated two homes for us, wrote in an email as follows:

> Sorry to disappoint you but on reflection we do not think we can write anything for the newsletter on the impact/effect of participating in the project. We were interested to take part, but not in anyway emotionally engaged. We look forward to reading the final report.

In worrying about 'what next' for our volunteers, we had inadvertently cast them as recipients rather than contributors. There was perhaps an implication in our request that they might have needed (and might still need) to volunteer rather than simply chosen to be involved in our research. This chimes with Baldock's finding regarding volunteering in the US that the social participation of 'seniors' is defined in terms of productive contribution to the community: an older volunteer is regarded as a 'healthy, autonomous citizen not in need of care' (1999, p. 599).

The nature of volunteering

Most participatory research in gerontology involves older people as representative service users, and their participation is both required and justified in terms of policy and practice needing to draw upon their views and experiences (Peace, 2002; Walker, 2007). Although our volunteers were older people, they were not representing older people in this way. Rather we would conceptualize them as 'amateur' researchers or 'researchers beyond the university walls' (Finnegan, 2005). They are amateur in the sense of not being employed to undertake research and not pursuing a professional career in research. Such researchers are, as Finnegan has pointed out, of paramount importance, in the fields of archaeology, astronomy and ornithology, and there is a long tradition of very high quality amateur research in the field of family and local history (Drake, 2005). We would argue that, through the tracing study, we tapped a rich seam and that gerontologists have much to gain by engaging with older people in this way.

Whether our co-researchers were recruited through an age-based organization, such as U3A, or an organization for retired professionals and academics, such as the Social Work History Network, the evidence suggests that most had a commitment to lifelong learning, and that following the ending of their employment careers, they have sought out opportunities (such as involvement in our project) to share and further their knowledge and develop new skills and expertise. In this, our project may be of relevance to recent debates in the emerging field of critical educational gerontology where the focus has shifted away from educational provision towards the experience of learners (Withnall, 2006).

It might be argued that the age of our volunteer researchers was of little significance. How different might it have been had we not specifically recruited older people? We would argue that their older age was significant for two reasons. First, they had reached a point in life where their relationship to paid employment was changing or changed. They were able to use their time differently and were open to new opportunities. Most of them were middle class and had been employed in what Parry and Taylor (2007) refer to as 'professional and creative' occupations as opposed to traditional working class or 'worker' jobs. These are the people who, come retirement, find it more difficult to relinquish employment and who are more likely to engage in voluntary work.

> There was a crucial difference between the trajectories of the 'workers' and the 'professionals and creatives' at state pension age in their differential involvement in voluntary work. Many of the latter drew considerable satisfaction from such engagement – for some, more than from their paid work. It offered an alternative strategy for continuing to gain 'occupational' satisfaction after state pension age, particularly for those unhappy with their inflexible paid-work arrangements. For the 'workers' however, these strategies were less relevant, because of their unfamiliarity with unpaid work ... and because they lacked the material security that facilitates such work.
>
> (Parry and Taylor, 2007, p. 594–5)

There is some evidence too that our volunteers may have been more receptive to the topic of our research than younger people, recognizing the importance of care homes in supporting older people with care needs in the community. Several made reference to their growing awareness and interest in ageing and later life and their increasing concern about care arrangements for themselves or their family. They may have expressed it in a self-deprecatory way, but it is evident that some were anxious that their mental abilities might decline or be perceived to be declining and that participation in this kind of activity would help retard the process.

Secondly, and perhaps more importantly however, the majority of the volunteers were at least teenagers or in many cases young adults at the time Townsend was conducting his research. Many, therefore, were his contemporaries. The late 1950s is a period of history through which they have lived and about which they have personal memories and knowledge, some of which relate specifically to residential care for older people and, for a few, to Townsend himself. Our volunteers were, therefore, not just amateur researchers but also people of an age who were researching a period of history of which they have first-hand recollection. For them, this was an opportunity to revisit aspects of their own earlier lives.

Regulation and quality

A further issue regarding our use of volunteer researchers relates to regulation and the quality of their contribution. As we explained in Chapter 3, there were no formal contractual arrangements with our volunteers. The process of self-selection was a potentially risky strategy and one that might raise questions from the professional academic community about the reliability of the data we received as well as possible ethical issues relating to the conduct of the research. The way in which we briefed and supported the volunteers was therefore extremely important.

After the tracing was completed, we rated each of the reports received from the volunteer researchers as strong, medium or weak according to whether it contained the factual information required for aggregated analysis. Four-fifths provided sufficiently reliable information for the analysis. The 'weak' reports were those where there was an absence of information about reasons for closure. Sometimes this was unavoidable, for example, where we were convinced that information was simply not available. There were a very few volunteers who gave up and their homes were reallocated to another volunteer or picked up by a member of the research team. Overall, however, the volunteers located and found out what happened to all 173 of the homes Townsend had visited, enabling us to address some key questions relating to the sustainability of residential care homes. As exemplified in Appendix 2, a majority of the reports had fully referenced sources and in a great many cases far exceeded our expectations in the quantity of information they provided. Many appended supplementary material which included copies of land registry documents, photographs and newspaper cuttings. This is perhaps a reflection of the kind of expertise and local knowledge upon which we had drawn.

Recruiting volunteer researchers is not straightforward and requires careful planning and preparation. Whatever the risks attached to such a strategy, the fact remains that the tracing study could not have been undertaken without the assistance of the volunteer researchers. They have created an invaluable historical record which will be deposited in the UKDA, one that will be available to researchers in the future.

Reusing data

Reuse involves going back to data that have already been analysed and from which conclusions have already been drawn and 'findings' published. As indicated in Chapter 3, there were some problems with the *The Last Refuge* data, both in the way they had been archived and in what was missing. Although, as Macleod and Thomson (2009, p. 127) suggest, these are 'at one level' practical matters, they are also 'profoundly linked to the making of history' and 'the stories' that can be told about the past. There was not a great deal of specific metadata in the *The Last Refuge* boxes although much contextual information was gained from the book, from archived interviews with Townsend and from Townsend in person. In discussing our reuse of Townsend's data, we have chosen to focus on three issues: disclosure, transformation and replication.

Disclosure issues

The use of archived data raises a number of ethical issues regarding confidentiality and disclosure. As we have pointed out earlier in this book, it is now a requirement of the ESRC that data from research it funds be offered to the UKDA. Consequently, our consent procedures included seeking agreement from our participants for the data we obtained through them to be deposited for educational and future research purposes. No such consent was obtained in the late 1950s when Townsend conducted his research for *The Last Refuge*. Indeed it is unlikely that, at that time, Townsend even envisaged his data being deposited in a national collection. It was only in 1994 that his data were deposited and we were fortunate that during the course of our research we were able to consult Townsend about disclosure issues. The ESDS Qualidata 'End User Licence Agreement' which we signed in order to obtain access to *The Last Refuge* boxes prohibits the disclosure of information that might identify the original research participants. However, it was because the homes Townsend visited had not been anonymized that we were able to design a project that began with a study that sought to trace the homes that he had visited. And in order for our volunteer researchers to trace these homes it was necessary to reveal some information about them. Given the possibility of a home still functioning as a care home, we were careful to supply the volunteer only with information that related to the location of the home. Nevertheless this raised issues for us about the right of our volunteer researchers to be party to the historical information to which we, as the accredited researchers, had access. For example, apart from reports on the homes, the archive contains photographs of some of the matrons and wardens of particular homes traced by our volunteers which would have been of interest to them but which we were not permitted to disclose.

Similar issues arose when we ourselves revisited homes that were still functioning as care homes. Understandably, some of the staff – who so generously were giving of their time and sharing information they had about the history of their homes – were curious to know what light we could shed on that history. Again this raised issues for us about how to handle the conflicting demands of research governance procedures which we had agreed to observe and the rights of research participants to knowledge about the places in which they lived or worked. For example, the proprietor of one of the homes we visited had, what we considered to be on the basis of Townsend's report, an inaccurate picture of the past. In this instance, we decided we should provide a précis of Townsend's report which omitted any judgements about the quality of the home at that time which might reflect on the owner in 1959. This omission was particularly important because the home had changed hands only twice since Townsend visited it and the couple who bought the home from the owner when Townsend visited lived in a house immediately adjacent the home and remained in contact with the present owners.

As with all social research, our dissemination activities, including the writing of this book, also raises disclosure issues. Central to our design for the follow-up study has been the need to draw on the original data on homes we have revisited and to quote them as evidence where appropriate in order to make comparisons. This poses questions about sourcing and traceability. And apart from protecting the identities of people and places involved in our research we are also aware of the need to respect the original researchers who have voluntarily opened their data for public scrutiny. In the end, as Bishop (2008) has pointed out, each case has to be assessed individually, and in addressing the ethical dilemmas surrounding disclosure the key questions are: what kind of harm might disclosure cause and to whom?

At the heart of these disclosure issues is what constitutes historical rather than sociological evidence. Are Townsend's reports now historical documents subject to different arrangements regarding reuse and disclosure? We cannot say 'yes' to this question because by bringing the past into the present through our research we have transformed *The Last Refuge* data as we discuss below.

Transformation

The term 'reuse' has been the subject of some debate. Moore (2007), drawing on evidence from other studies which have used archived data,[1] argues that it is not the original data which are being reused. Rather, she suggests, through the process of 'recontextualization' the data have been transformed into something different. So in what ways have we recontextualized Townsend's data? As researchers located in a different temporal and socio-cultural context, who are what might be termed late career researchers (unlike Townsend and Pinker who were in the early stages of their academic careers), we have viewed the archived data through a very different lens. Secondly, as mentioned above, we have relocated the data into a different temporal frame which changes its status and its nature. We now know, for example, that some of the 1958–9 home reports relate to homes that are still functioning as care homes while others relate to homes that are closed and may even have been demolished. Furthermore our own data amplify the original data through the acquisition of additional information about the history of the homes.

Most research which is based on the use of archived data has produced new data by asking new questions of the old data. Our research questions related to continuity and change and the very process of data transformation is integral to our understanding of this. Our analysis of Townsend's photographs provides a good example.

We have argued elsewhere (Rolph et al., 2009) that the 38 photographs Townsend published were strategically selected, arranged and placed in the centre of his book so as to conduct an argument. As we have explained earlier in this book, one of Townsend's key concerns was his finding that the majority of older people were still being accommodated in former workhouses which, he argued, should be demolished or used for other purposes. He put forward a powerful argument about the impact of these institutions on people's lives. Furthermore he highlighted the gross inequalities in provision and demonstrated the connection between social class and different types of homes.

In analysing the photographs he selected for his book, we identified a number of subject groups: exteriors, entrances and exits, staff, and residents. Photographs of the exteriors we have argued are used to draw a contrast between the vast scale, isolation and prison-like appearance of an ex-PAI in East Anglia and three post-war local authority homes: two elegant urban converted homes set in well-kept grounds containing flower beds and trees, and a modern purpose-built home. The photographs of entrances and exits also emphasize size and scale, remoteness, bleak undecorated corridors and archways, empty of people and comfort. These images tell a story of hopelessness and oppression, their central and dramatic vanishing points and deliberate perspectives offering a metaphor for another endpoint, death, and symbolizing the seeming inevitability of these 'last refuges'. The photographs of staff again draw contrasts, this time between the uniformed staff and apparently forbidding matrons of both the former workhouses and other local authority homes and the warmth and friendliness of a matron posed with a resident in a voluntary home.

The most substantial group of photographs, that of residents, includes scenes of residents engaged in domestic work, of sleeping arrangements and eating arrangements and other aspects of daily life. Again these photographs illustrate marked contrasts between the harsh conditions in the ex-PAIs and the more favourable conditions in the smaller converted homes. As we describe in detail elsewhere (Rolph et al., 2009), they illustrate class differences through the clothes the residents are wearing, the type of activities they are engaged in and the facilities they have at their disposal.

Townsend includes 31 captions to accompany these photographs and these we have argued, in line with Hall (1973) and Chaplin (2006), lead the viewer towards a particular meaning and interpretation so that we see what Townsend wanted us to see above what else there is to be seen. For example, 'Dining-hall in a former workhouse, *with recess used as dormitory for 56 men*', 'Day-room *with no floor-covering* in a former workhouse' (our italics) highlight the unacceptable. Supporting his argument about institutionalization, 'Apathy in a day-room', 'Mid-morning in one institution' and 'Mid-afternoon in another', all imply the endless days of lethargy in gloomy day rooms where there are no means of stimulation and people are simply, as Townsend has said, 'confined to quarters' (Townsend, 2007). His argument becomes even clearer when the photographs are linked to the text. Although the links are not explicit, they are not hard to detect. For example, the following description of life in an unspecified day room in a former workhouse seems to chime with the photograph 'Apathy in a day room' (Townsend, 1962, Illustration 30) which depicts one man slumped over a table and another being raised up by an attendant.

> Half the men seemed always to be asleep, some of the others were staring at the windows, and about four or five were reading newspapers. ... A woman attendant went up to one man slumped with his head between his knees. 'Lift yourself up, love, you'll fall and crack your head open'. 'A bloody good job if I do.'
>
> (Townsend, 1962, p. 105)

When the photographs are viewed alongside the text in this way, both visual and textual meanings are enhanced.

Townsend portrays many of the people in the former workhouses as inactive and dependent, and, by implication, on the margins of existence. There is pathos in his photographs and the reader is led to feel indignant on their behalf. Emphasizing pathos in this way can be construed as disempowering his subjects. In exchanges with us Townsend disputed this interpretation and argued that some of the photographs also illustrated what he called in the text 'a passive form of hostility' to the life in the homes (Townsend, 1962, p. 105). This indicates the complexity of interpretation, as Rose (2007) has pointed out, and that different audiences at different times enable both pathos and resistance to be read into these images.

Our analysis of Townsend's photographs is just one illustration of how original data can be transformed into something different. Townsend never treated his photographs as 'data'. They were there, as he subsequently explained to us, to tell a 'complementary story'. Fifty years later, we see these photographs very much as data open to interpretation – data we decided to analyse and to compare with our own photographs taken in the homes we revisited.

Replication

For comparative purposes, we decided to replicate Townsend's methods and this included the taking of photographs. In Chapter 3, we described in some detail our approach to replicating Townsend's research instruments and how the changes we had to make to his schedules are themselves evidence of change. Townsend did not include the taking of photographs as part of his methodology (see Townsend, 1962, pp. 13–14). However, as we have argued above, his photographs have become an important source of data, hence our own interest in repeating this aspect of his research. Replicating his photography raised some further methodological issues for us in relation to replication. Changes in technology, as well as new research governance procedures and changed ethical practices meant that what we were able to photograph was very different to what Townsend was able to do. These changes had a significant impact on our photographic portrayal of life in the care homes we visited and of those who live and work in them and therefore on the kinds of comparisons we drew between the late 1950s and the early twenty-first century.

Some photographers have attempted to document social change through 're-photography' (Ganzel, 1984; Rieger, 1996; Harper, 2001). This was not our intention. Viewed as specific socio-cultural products, to try to recreate Townsend's images using his precise viewpoints, angles or poses would have been an impossible aim. Building on Davies and Charles' (2002) line of argument (see Chapter 3), the necessary changes we had to make to our own use of the camera contributed to our understanding of continuity and change in the homes we revisited. And these changes related to both composition and process.

Our purpose in using the camera was to generate photographic data which could be used to make some historical comparisons with Townsend's photographs. Decisions about what to photograph were dictated both by Townsend's subject groups and by scenes within the homes that caught our eye and mirrored our own observations regarding residential care in 2005–6.

Of relevance to our own use of photography was the kind of camera we should use: whether, like Townsend, to use a Leica camera to produce black and white photographs or to take advantage of the latest technology to produce digital colour or monochrome images. There are debates in the anthropological literature about the technical choices open to researchers. For example, some suggest that black and white photographs create an atmosphere and in fact reproduce more effectively the 'new reality' created by the photographer. Others suggest using a combination, as 'different meanings can be elicited from the use of black and white and colour within ethnographic photography' (Canal, 2004, p. 33). Canal suggests the use of both in historical research, which is what we decided to do. The upbeat appearance of our colour photographs is undoubted and often does not coincide with our memories or field notes concerning a particular home. By comparing them with our monochrome versions of the same scenes, we were able to achieve a more balanced view. The colour photographs, like the recently discovered colour photographs of the Depression (Hendrickson, 2007), can warn us against adopting too gloomy an interpretation. They enable a perspective which moves away from one that is in our mind's eye, habitually black and white (following Townsend), to one that is still capable of shocking, but also in many cases creates a more positive effect (ibid., p. 13).

While the technical choices available to us opened up new opportunities, changes in research governance procedures, particularly in relation to informed consent, presented us with challenges not always presented to Townsend. His research was funded by the Nuffield Foundation and was, therefore, independent of central government. He has pointed out how important this was and how useful in gaining the confidence and co-operation of the participating homes. He was able to offer assurances to the wardens and matrons he interviewed that nothing would go to 'the council', and that his findings would be generalized and anonymized (Townsend, 2005). It is perhaps important to realize that his research was the first major study ever conducted of residential care for older people in Britain and that academic researchers, particularly in the social sciences, were much more of a rarity than they are today. As such he carried considerable authority, and many of the homes, particularly the local authority ones, were simply told by the chief welfare officer for the county that they would be co-operating. Consequently, he was able to take photographs in a way that is impossible today.

Our research has had to be carefully negotiated with each home, most of which are no longer publicly owned. In particular, as we have described in Chapter 3, our proposed procedures regarding consent had to be approved by an Ethics Committee and we, the researchers, had to obtain Criminal Record Bureau clearance in order to conduct interviews within a care home. Our consent forms, both for individual residents and for the home manager included permissions relating to photographs. Using a digital camera, we were able to show the portrait photographs we had taken to residents and staff and to delete any they were not happy with. However, obtaining informed consent from care home residents is not always straightforward. Many residents today are more physically and mentally frail than those Townsend encountered, and many have dementia. In some cases gaining written consent required a witness or advocate.

The potentially negative impact of these kinds of procedures on the quality of data produced in relation to vulnerable groups has been discussed by Crow et al. (2006). For us it meant that the business of taking photographs was constrained. As a specific example, how could we ensure, if we wanted to take a photograph in a lounge or other public space, that we had everybody's consent not only to take the photograph but also to publish it? Some people may have been asleep; others not aware of the implications of their agreement.

There were also issues about intruding on people's privacy. Some scenes photographed by Townsend, such as one taken in the men's washroom, would be regarded as inappropriate today. We also encountered occasionally what we regarded as unacceptable practices such as a resident having her hair dried beneath a hood hairdryer in the middle of the lounge, surrounded by other residents and guests. Photographs of such scenes might provide important evidence of poor practices, but to produce such evidence would be a violation of dignity. As a consequence our photographs, by comparison with Townsend's, arguably present more empowering images which lack the pathos of some scenes that we witnessed.

Instead of the large day room scenes taken by Townsend, which capture the 'block treatment' aspects of care home living, our photographs feature small groups of residents, or are portraits of individuals, all of whom granted permission to be photographed. Some of our portraits bring out a dignity which is perhaps indicative of resistance and struggle. Photographs are a sensitive issue for many people and can be potentially distressing particularly when they reveal frailty, disability and decline and sometimes residents and staff exercised a preference not to be photographed. Our approach, therefore, meant engaging with our subjects in a way that Townsend may not have done. A resident's letter to one of us was testimony to this:

> I wish to give you a very big 'thank you' for the magnificent picture of myself you sent me, following our interview ...
>
> I am very pleased with it, and shown it to members of the staff as well as residents. It has so much detail to it, showing clearly the actual time it was taken, by the watch I was wearing, and also clear pictures of my wife and our wedding, and other interesting details of importance to me.
>
> I trust the interview was as satisfactory to you as it was to me, and thank you for sending a copy of the photo to me. It is now hung on the wall with other pictures I am proud of.

The man who wrote this letter also gave us permission to publish the photograph indicating the importance of recognition and acknowledgement as well as protection (see Photograph 9.1). In more recent years, there has been substantial debate about the issue of anonymity versus ownership

Photograph 9.1   Reg Paine, 2005
Source: *The Last Refuge* revisited collection. © Randall Smith

(Gluck and Patai, 1991; Finnegan, 1992; Yow, 1994; Summerfield, 1998; Walmsley, 1998; Rolph, 2000). While there was a need to guarantee anonymity so that participants were able to entrust us with their views, there also needed to be the opportunity for people's contribution to be valued and acknowledged. We designed our consent forms for both residents and managers so as to offer them the opportunity for personal acknowledgement. Appendix 1 is testimony to the importance of this and also supports the argument that social scientists need to develop an ethics of recognition and not just an ethics of protection (Sweetman, 2009).

Conclusion

We have focused on two very specific aspects of our research design – involving older people in research and reusing data – both of which are very topical in the early twenty-first century.

Older people have participated in our research both as researchers and the researched. Comparing our relationship with our participants and Townsend's with his might suggest some changes. For example, it might be argued that the residents of the homes we visited have been transformed from passive welfare recipients to active service users or consumers. However, we concluded that such familiar rhetoric is an oversimplification. When we began to contextualize our respective relationships, it became clear that the way the older people have been represented both in Townsend's research and our own is complex and open to interpretation. As McLeod and Thomson argue, context

> encompasses local, national and global levels, theoretical and substantive concerns that were urgent at the time of doing the research, and the dynamic of the research encounter.
>
> (2009, p. 135)

What we have argued in this chapter is that the contrasting images of care home residents that have been created by the research of Townsend and ourselves reflect differing concerns and the extremely different dynamics of the research encounter, rather than a contrast between victims and beneficiaries.

Reusing archived data alerts one to the practical and ethical issues associated with data deposit. At the time of writing this book, we were preparing our own data for archiving. Knowing that our data might be archived and perhaps reused in the future undoubtedly had some influence on its production, and this may have involved an element of inhibition on our part. It was always our intention to archive our data for use by future historians and to this end, while guaranteeing anonymity to our research participants, except where they wished to be acknowledged, we obtained consent from the participating homes to retain their identities for future research purposes. In this sense, we see ourselves as creating an essentially historical record but one derived from qualitative social research and governed by current ethical procedures. Our research has been very much on the border between these two disciplines.

In this chapter we hope to have demonstrated that revisiting studies, such as ours, offer the opportunity to examine many facets of continuity and change relating to both the 'what' and the 'how' of research. In the next and final chapter, we focus on our conclusions about what has happened to residential care for older people over the last 50 years.

10
Continuity and Change in Residential Care for Older People

We begin this chapter by summarizing our findings regarding residential care for older people over the last 50 years. Has nothing changed, as is often alleged? Or has residential care been transformed? We go on to consider Townsend's policy recommendations and how these tie in with the policy context at the time of writing this book in 2009. In so doing we consider briefly what our research contributes to current debates relating to the theorization of old age.

Stability and change

Historians often remind us that we cannot fully understand the present without an understanding of the past. Those familiar with current social policies for older people and the provision of residential care in particular, know of the changes that have taken place. Most notably perhaps is the decline in public sector provision, the rise of the private sector and changes in the way homes are regulated and inspected. There are also the changes in the demographic characteristics of residents and in the way that homes are staffed. All these changes are reflected in our findings and perhaps come as no surprise. What our research reveals, however, is how such trends have affected the fortunes of individual homes. Furthermore our detailed comparison between then and now demonstrates the scale of change which is easy to underestimate.

Our research shows that ownership of nearly all the surviving ex-PAIs and other local authority homes (except those in Wales) had been transferred to the private and voluntary sectors. Reflecting the trend towards corporatization,[1] several now belong to large non-profit trusts or private for-profit companies. This trend does of course raise the question of what kinds of homes the large corporate providers are investing in. During the course of the tracing study, we came across some examples where the old site of a home visited by Townsend was being redeveloped to include large new-build care homes, such as the 90-bedded home described in Chapter 3 in the grounds of St Mark's hospital. Townsend was particularly concerned with the issue of size and its negative impact on the lives of residents. Nearly 50 years later, Parker et al. (2004) demonstrated a direct negative association between large-scale homes and the quality of life of residents. The economies of scale that appear to be characterizing new developments might be regarded with some concern therefore.

Among the homes we visited in 2005–6, we found that much more personal care was required by residents than was the case in 1958–9 and that the numbers of care staff employed were far greater, although many worked part-time hours. Residential care, when compared with 1958–9, has become a predominantly female domain where a largely female workforce is caring for a largely female population of residents. We found that staff roles were more differentiated and that proprietors and managers were doing far less 'hands on' work and much more administration when compared with the past. This in part reflects the much closer regulation and inspection of care homes in 2005–6. Overall, staff in 2005–6 were better trained, albeit still inadequately. While there is less reliance on casual and familial labour, staff remain poorly paid and often are not protected by union membership.

Rather than segregation according to sex, the homes today are more likely to be segregated according to perceived functional ability. Several have discrete nursing units and dementia care units and have taken on functions that in the past belonged to the NHS. This seems to reflect the shifting balance between residential care and nursing care places in care homes. As Table 2.1 shows (see Chapter 2), between 1990 and 1995 the ratio of nursing home places to residential care places increased substantially from 38 to 68 places per 100 residential care places, a ratio that has since remained fairly stable. We have concluded that residential care for older people, insofar as it is now catering for an older and more infirm population, has been transformed into a radically different instrument of social policy when compared with the 1950s.

In assessing the quality of provision using Townsend's measure, the standard of care has improved substantially, particularly in the physical facilities available to staff and residents. However, in applying his measure to the homes we visited, and in comparing it with current measures, it is clear that the goal posts have shifted: standards and expectations have changed significantly. While the physical environment may have improved, we recorded many institutional features which appear to characterize the twenty-first century care home. These features reflect not only the changed function of residential care but also an increasing concern with risk and safety. The harsh attitudes that often characterized the workhouse regimes may have disappeared but have been replaced by what Parker et al. have referred to as a 'risk averse' culture which they found threatened the quality of life of residents, particularly the least frail residents. As they point out:

> The *National Service Framework for Older People* (Department of Health 2001) endorses the notion that older people should be able to 'determine the level of personal risk they are prepared to take when making decisions about their own health and circumstances' but it appears that the directive has as yet had little impact on the care culture of residential settings.
>
> (2004, p. 958)

All these changes are what we might have expected given the shifts in policy and practice over the years, together with the trend towards the 'risk society' and the individualization of risk (Beck, 1992; Giddens, 1994; O'Rand, 2006). What are less familiar are the continuities, and this is where our longitudinal approach has provided some new, and perhaps unexpected, perspectives on the history of residential care. The tracing study, for example, revealed that the former voluntary homes had the highest survival rate. One in three of them were still functioning in 2005–6 and 12 of the 14 were still owned and run by the same charity as in 1958–9. These included some large organizations as well as individual homes arising from a single bequest. As detailed in Chapter 4, some of these homes had collections of documents and photographs detailing a rich history. Our findings indicate a stability in this sector. In a context where home closures, particularly in the public sector and parts of the private sector, have been increasing in recent years, such stability is important because care home residents have no security of tenure (Johnson, 2002).[2] Furthermore, the quality of the voluntary homes, in keeping with Townsend's findings, exceeded that of other tenures despite the fact that the fees were sometimes considerably lower than those in the private for-profit sector. Our findings also suggest that where these voluntary sector homes had a strong religious or secular identity they were able to create a community that had the potential to meet residents' social and emotional as well as physical needs.

In addition to the voluntary homes being of an overall higher quality than the private ones, a further continuity with the past was the uneven quality of provision, particularly within the private sector. Townsend's reports illustrate the differences to be found between the ex-PAIs at one extreme and some of the smaller local authority, voluntary or private homes at the other, as well as among the private homes he visited. As Chapter 8 indicates, we found similar differences, despite the introduction of National Minimum Standards at the beginning of the twenty-first century, indicating as Townsend did nearly 50 years ago, 'the different living standards enjoyed by different people with similar needs' (1962, p. 432). This is evidence of the persistence of structural inequalities which limit the exercise of individual agency and supportive of the arguments of critical gerontologists such as Walker (2009) in the face of criticisms from cultural theorists (see, for example, Gilleard and Higgs, 2000, 2005).

There have been times when academic colleagues who have some familiarity with *The Last Refuge* have expressed surprise that any of the homes Townsend visited are still functioning. There is a tendency for people to forget that the ex-PAIs were only a quarter of Townsend's sample which included many other types of homes: other local authority homes, voluntary and private homes. This is perhaps because Townsend conveyed such a forceful message about the appalling conditions in many of the ex-PAIs where large numbers of older people were accommodated at the time. His message was strongly supported by his photographs which, as we have argued, have left a lasting image of residential care in many people's minds (Rolph et al., 2009).

The negative image of residential care is perhaps the most striking continuity of all. As we reported in Chapters 6 and 7, we found some residents adopting a variety of strategies to distance themselves from other residents and the stigma attached to being a care home resident. For many, residential care had been an option of last resort. As we have argued in Chapter 2, this negative view of residential care has been a subtext of policymakers for several decades and has become particularly prominent in more recent policy developments in the field of social care.

Social care in the twenty-first century

In discussing the development of future policy, Townsend remarked that 'a more positive policy might start with sheltered and special housing' (Townsend, 1962, p. 399) and recommended that 'in the long run such housing should largely replace residential homes' (p. 404). He also suggested that improved domiciliary support services would in large part negate the need for residential care. Although residential care has not been abandoned as he had hoped, recent policies have been undermining it by promoting alternative strategies for meeting older people's needs and preferences. Two developments in particular are 'personalization' and 'extra care' housing.

Part of the strategy to develop personalized services, as we discussed in Chapter 2, included the use of direct payments and individual budgets. Individual budgets were particularly promoted by 'In Control', an organization representing the interests of people with learning difficulties. Its model of self-directed support (which acknowledged the issue of finite resources) began in six local authorities in 2003. The Department of Health commissioned 13 two-year pilot studies on individual budgets, starting in March 2006.

In advance of a promised Green Paper[3] on the future funding of social care, the government published what it called a historic protocol in December 2007. This argued that there was an urgent need to develop 'a new adult care system' based on the government's commitment 'to independent living for all adults' (HM Government, 2007a, p. 1). It comes as no surprise that there was no explicit reference to residential care in the text of this essentially aspirational document. The protocol was swiftly followed, in January 2008, by a Department of Health circular, *Transforming Social Care* (DH, 2008), which again emphasized personalization, early intervention and prevention. Everybody eligible for social care support would have a personal budget, 'to enable them to make informed choices about how best to meet their needs' (p. 5). Such choices should not, it stated, be confined to those living in their own homes. The need to reduce spending on social care was emphasized, and in particular it was noted that residential care was 'by far the biggest area of expenditure' (Lombard, 2008).

In May 2008, the government published a consultation document in which the prime minister said it was important to make it easier for people to be cared for in their own homes 'so that far fewer people have to move out'. Yet again, the focus was on reducing the demand for residential care, accompanied by greater emphasis on support from family members and friends (Number10.gov.uk, 2008). Although the future of residential care was not prominent in the text, the familiar point was made that people want to stay in their own home 'and avoid institutionalisation' (HM Government, 2008, p. 14). The current social care system was criticized as having 'a tendency to create over-reliance on residential care' (ibid.).

Extra care housing has become 'a key plank of government policy in terms of its aims to promote choice, independence and well-being for older people' (Evans and Vallelly, 2007, p. 8). In the first decade of the twenty-first century, through the availability of various funding streams in England and Wales, there has been a surge in the development of extra care housing. The history of sheltered housing and other specialist and supported housing options, however, has not been an undiluted success (Tinker et al., 1995; Audit Commission, 1998). In the 1970s and 1980s, very sheltered housing schemes emerged which offered more support than ordinary sheltered schemes but which were not residential care homes under Part III of the 1948 National Assistance Act (Means and Smith, 1994, p. 184). Wright et al. (2009) describe very sheltered housing as the forerunner of extra care schemes. Citing Croucher et al. (2006), they suggest that extra care housing can be viewed either as a remodelling response to the inadequacies of existing sheltered housing schemes and care homes, as an alternative to residential care, or as an innovative positive response to ideas about promoting the independence of service users, as reflected in initiatives such as direct payments and user influence on the provision of services. Extra care housing, it has been claimed, is not just about the quality of care but about the quality of life (Riseborough and Fletcher, 2008) and there is no doubt that many older people with care needs see it as a preferable option to a care home. As Wright et al. (2009) point out, in contrast to care homes, extra care housing gives residents greater security of tenure and provides self-contained accommodation.

One of Townsend's major criticisms of residential care was that it was a segregated form of provision and one that was often cut off from the local community. While extra care housing may offer more independence for its residents, it remains an age-segregated facility, as Evans has pointed out:

> The Extra Care housing model shares the age segregated characteristics of residential care. ... On the positive side many extra care schemes are well integrated with their local communities and are increasingly acting as hubs for local health and social care services. Many also have restaurants, shops and other facilities that can be used by local residents. Location is key in this respect, with most schemes being physically integrated with existing housing, be they new build or reprovision of residential care. To this extent they have many advantages over retirement villages, which tend to be fairly remote due to the cost of land and planning restrictions. In terms of size, extra care schemes tend to be reasonably compact, with up to 100 units of accommodation. However, the distinction between

extra care schemes and retirement villages is becoming increasingly blurred and there are now some extra care 'villages'.

(Evans, 2009)

Although Evans describes 100 units as 'reasonably compact', when compared to the size of the care homes we visited, this is large, and also large by Townsend's standards. During the course of our research, one of us visited a new extra care housing development which is to replace one of Townsend's surviving homes (now offering intermediate and respite care only). Its scale was particularly striking, resembling something like a university student accommodation block. We might ask therefore, are these schemes the new segregated institutions of the twenty-first century?

Behind both personalization and extra care housing there appear to be two key policy drivers. The first is that within the context of an ageing population, the costs of institutional care are too high. The second is that residential care is not the preferred choice of older people.

Regarding costs, while it could be argued that the cost to the resident might be less in extra care housing, it is important to remember that the lower cost of providing extra care housing compared with care homes is closely linked to what is defined as '24-hour care'. Wright et al. note that 'Night care may be minimal' (2009, p. 138) so, should it be needed, the costs of not having blanket 24-hour care coverage will be borne by other services, such as the NHS. And costs more generally, monetary and non-monetary, may fall on the individual resident and his or her family, friends or neighbours. Likewise, the use of individual budgets and direct payments could similarly transfer costs, particularly now that direct payments can be used to pay a family member for care.

On the issue of choice and preferences, we would not contest that most older people wish to remain in their own homes if possible. However, an independent review of the individual budget pilots found that older people were less likely than others to report higher aspirations regarding their use (Glendinning et al., 2008). Many of the older people (or their proxies) interviewed, who were supported by adult services, indicated that they did not want the 'additional burden' of planning and managing their own support, and a similar view was expressed by care providers and co-ordinators. While this might be a reflection of low expectations on the part of older people themselves and their supporters, anticipated drawbacks included the responsibilities of managing a budget, making new support arrangements and the risks associated with directly employing care workers rather than having them provided by an agency or local authority. The authors concluded that it might take some time for older people to develop the confidence to assume greater control over their care arrangements (ibid., p. 19). In the meantime, as Stephen Burke, chief executive of Counsel and Care, has commented, 'for people with dementia or near end of life, personalisation is still more rhetoric than reality' (McNabb, 2008).

Furthermore, the policy rhetoric about 'institutionalization' has tended to overlook the evidence – evidence that we have also found – that reflects positive (as well as more negative) views about residential care. For example, in the mid-1960s, an official survey of social services for older people in 13 local authority areas concluded that a high proportion of the people in residential care were satisfied with their home (Harris, 1968). An overview of studies on the demand for residential care concluded that 'a minority (about a fifth in each case) said that they would be pleased or very pleased to enter a local authority residential home, private rest home or nursing home' (Sinclair, 1988, p. 248). A 1990 survey reported that 'the idea of residential care was not as abhorrent to the elderly people living in the community as might have been imagined'. A third of respondents said they would consider it (Allen et al., 1990, p. 105). Of those already living in a care home, 19 per cent said they had asked to come into residential care, a proportion very similar to that reported by Sinclair. More recently, an ICM poll revealed that people between the ages of 35 and 64 were more afraid of the prospect of moving into a care home (72 per cent) than those aged 65 or more (56 per cent) (Carvel, 2007). These findings suggest that older people develop a more nuanced view of ageing and residential options for later life.

As our research has shown, people move to live in a care home for a variety of reasons and for some it is a positive choice. For others, given their situation, it may quite simply be the best of a limited range of options. It has been estimated, for example, that 'over a third of people with dementia live in care homes and that two-thirds of care home residents have dementia' (Manthorpe, 2008, p. 36). Extra care housing has been found to be a popular option for older people with dementia and their families and one that can provide a good quality of life. However, when residents become more frail and require high levels of care, pressure is put on resources (Vallelly et al., 2006; Evans and Vallelly, 2007, p. 9; Dutton, 2009).

Darton et al. (2008, p. 4) reported that in 2007 there were about 37,600 extra care dwellings in England, compared with about 270,000 personal care places and 449,000 personal and nursing care places in care homes in the UK. According to the January 2009 fourth and final CSCI report there were in England, in March 2007, 250,000 adults receiving residential and nursing care paid for by local authorities, of whom 191,500 were aged 65 or over (CSCI, 2009, Table B1, p. 168). With the best will in the world, changing the balance between 37,600 extra care dwellings and 191,500 supported residents in care homes for older people in England would be a major challenge for any fundamental reform of the funding of long-term care.

Furthermore, there is evidence that private investors in adult care (bankers and venture capitalists) prefer less risky traditional models of care, such as care homes, rather than more personalized services, such as extra care. In 2007 it was reported that

[i]n the past year private equity giants such as Blackstone Group and 3i have made huge cash windfalls through selling their elderly care businesses. Eager new investors such as the Qatari royal family and legendary Irish tycoons John Magnier and JP McManus have been quick to replace them. Investors see a ripe market for consolidation. The biggest player is quoted firm Southern Cross with 36,000 beds. It owns just 7 per cent of the market but has ambitions to treble that. ... City analysts say the boom will continue. An expanding elderly population and a supply of cheap eastern European non-unionized labour is fuelling growth and flattening costs.

(Mathiason, 2007)

As pointed out in Chapter 4, the boom did not last for Southern Cross and in 2008 it ran into financial difficulties.[4] Nevertheless what drives investment in social care cannot be ignored.

A similar picture emerges when we look at the take-up of direct payments and the use of individual budgets. In March 2007, about 54,000 people were using direct payments in England, and in September 2007 about 1000 people had individual budgets (CSCI, 2008, p. vi). The total of the latter was projected to reach 25,000 by June 2009 (In Control Newsletter, Issue 5, May–June 2008, p. 1). By way of contrast, the total number of places in residential homes for adults of all ages was 263,000 in 14,500 registered care homes (CSCI, 2008, p. 169). So the take-up of direct payments and the extension of individual budgets have a long way to go before they exceed or replace the provision of residential care.

The commitment to independent living for all adults may be a popular policy objective but it allows, or even encourages a denial of what several long-standing researchers have revealed to be the reality of the lives of many older people in care homes. Clough (1981, 1982, 1996), for example, has pointed out, 'ageing is not always easy: there are those who are weary, sad, depressed, confused, who want to be looked after and do not want to take decisions' (Clough, 2008, p. 6). Victor similarly argues that residential care should remain an important element in the state's response to the 'problems' of ageing and that it has been underplayed in recent policy documents. 'What are required are appropriate responses to the problems faced by dependent elders, not predetermined ideologically based positions on the supposed merits or disadvantages of specific types of care solution' (Victor, 2005, p. 319).

De-stigmatizing residential care

On the basis of our own findings, we would agree with the argument that residential care has a role to play in care provisions for older people. Nevertheless we would take issue with the emphasis on the 'problems' of ageing and, in particular, with the way they are associated with residential care. This is not to deny the very real challenges that old age can entail, but labels such as 'dependent elders' are stigmatizing and when linked to residential care, stigmatize it as a resource.

In July 2009 an article appeared about 'life in an old people's home' in *The Guardian* newspaper headlined:

> The staff here are kind, the rooms are bright, yet none of this compensates for the home's unspoken function: a place where elderly people are left by their families to die.
>
> (Gentleman, 2009)

This article seemed to us to reflect and sustain a number of assumptions: first that all older people have 'family' who are able to care for them when the need arises, second that those who do not fulfil this duty are failing in their moral obligations, thirdly that living in a care home means you have been abandoned and fourthly that moving to live in a care home is never a positive choice. Our research has demonstrated that these are not safe assumptions to make. Furthermore, it might be argued that what people fear most is not residential care per se but ageing and the challenges of deep old age or what Gilleard and Higgs have theorized as 'the fourth age': 'The fear of the fourth age is the fear of passing beyond any possibility of agency, human intimacy or social exchange' (2007, p. 8). Regarding care homes, their description of the prospect is apocalyptic:

> Ending one's days in a nursing home has become a prospect for many reaching 65, but it is no longer seen as the consequence of improvidence or indigence but of a totalizing infirmity pared of other significance. No longer the fate of the poor or a choice to be considered by the leisured classes, the nursing home has become a new space – a new void – within society. [...] it remains every bit as terrifying as the workhouse.
>
> (2007, p. 10)

Assuming the prospect of entry into a home to be universal and irreversible, they conclude that 'the nursing home now has a significance scarcely possessed some 25, let alone 50 years ago' (ibid.).

Whether one agrees with this argument or not, what it suggests is that care home provision is perceived as a homogeneous and universally negative category. The reality we found was great diversity, not just in terms of uneven quality but in regard to the character and culture of individual homes. The homes we revisited have very distinct histories stretching back 50 years and more, offering a rich variety of provision. Likewise the older people we met did not constitute a homogeneous group: some were very frail, and some were depressed or confused, but by no means all. Furthermore, as we have shown in this book, to theorize these people as 'ageing without agency' is we suggest a gross oversimplification. As Walker (2009) argues, the cultural theorists (such as Gilleard and Higgs) ignore the structural factors that generate variations in the exercise of individual autonomy.

In this book, we have presented overall a more positive view of residential care than that presented by Townsend. In Chapter 1, we quoted from Kathleen Jones' review of *The Last Refuge*:

> Old people, like young people, can be highly ambivalent, and their actual field of choice (as opposed to their ideal field of choice) can be very restricted. ... This is not really a book about old people, for their attitudes, emotions and reactions are judged entirely by middle class, young middle aged standards. It is a book about the author's belief that institutional living is undesirable.
>
> (Jones, 1962, p. 31)

We can also be accused of judging the lives of older people from a middle-class perspective but, unlike Townsend at the time, all three of us are much closer in age to those whose lives we have been observing. Perhaps we recognize more clearly than he did the ambivalence to which Jones refers. On the other hand, our research has demonstrated that the older people accommodated in the homes today are on the whole rather different to the occupants of 50 years ago and their feelings of ambivalence about living in a care home may be more acute.

To conclude, we would like to make a simple contrast. Peter Townsend started with a blank sheet and the resources to undertake and analyse a nationwide survey of residential care and to analyse the data he obtained from a strictly random sample of homes. As pointed out in Chapter 1, he also produced a fund of illuminating but impressionistic qualitative data (Crossman, 1962). We, in revisiting his study, decided to face the challenge of finding out what had become of the 173 sampled homes, and then to match his surviving, largely qualitative, data on surviving homes. What we have not done – nor was it our intention to do – is a complete restudy of residential care for older people in England and Wales. In our view, given the wealth of data and research on residential care over recent years (although having said that, there is surprisingly little data available at the national level), what was needed was research that focused on what becomes of care homes and how they deliver care as time goes by, week by week and year by year. This is what we hope we have achieved. We have compiled some unique historical data which will be archived for the use of future researchers, in the hope that it will contribute to policy that is informed by history as well as by gerontology.

Appendix 1

We would like to thank the following for their participation in our research:

Winifred Aitkens
Marian Allen
Francesca Ashburner
Arthur Ayshford
Joan Baraclough
George Bathurst
Irene Bayley
Joy Bell
Glynis Benford
Jean Bogle
Yvonne Bonas
Elaine Brace
Susan Bradbury
Eric Braim
Helen Bramley
Margaret Briggs
Elizabeth Brittain
Majorie Brooks
Rosemary Bryden
Mary Caller
Barry Chalkley
Susan Chambers
Susan Cheetham
Hilda Collier
Pat Constable
Patricia Cook
Margaret Cosh
Morag Crawley
Mick Curran
Cyril Dainow
Dave Davis
Jane Dixey
Jean Drew
Kim Drury
Lilian Evans
William Evans
David Fenton
Maureen Fisher
David Fox
Frances Fox

Eileen Fletcher
Tony Gardiner
Ian Gates
Jean Gibbs
Jane Gorman
Bert Green
Tony Gregory
Rose-Marie Grimwade
Naomi Gush
Liz Harding
Alison Hargreaves
Judy Hannon
Mary Harrison
Barbara Hawkes
Peggy Heeks
Sylvia Hewitson
Jean Hicks
Tina Hicks
Ruby Hoare
Joy Hollamby
Elaine Holder
Bill Howson
Pam Hughes
Claire Hunt
Jane Hydon
Doreen Ingram
Maggie Jesson
Greta Jolliffe
Joan Kemp
Heather Kent
Linda Kourellias
Grace Kowala
Gina Langford
Harry Lawford
Brian Lillington
Dick Loverage
Margaret Lyndley
Marcus Lyward
Margaret Macpherson
Patricia Martin

Lyn Meehan
Ellie Milsom
Campion Mead
Olive Medlam
Ruth Meltz
Daksha Mistry
Lilian Morgan
Margery Morgan
Elizabeth Noller
Jean Owers
Pat O'Connor
Bridget Ogden
Bernard Openshaw
Marie Ormerod
Brian Padgham
Alison Pendleton
Doris Pigney
Nellie Pink
Joy Pollock
Adeyinke Popoola
Terry Powell
Barbara Prynn
John Rainey
Philip Rapaport
Barbara Ratcliffe
Irene Richards
Norman Richards
Pauline Richards
Beryl Rumgay
Derek Seaton
Edith Senior
Ann Shaw
Marie Shell
Jean Shipley
Moyra Sidell
Liz Simpson
Ron Smith
Eileen Smithers
Ted Solilak
Marion Sottilini

Julia Spruntulis	Liz Van der Vord	Ronald Watkins
Dorothy Sweet	Winifred Venner	Claire Wendelken
Muriel Tate	Peter Wade	Stephen Wetherall
Robert Tate	Eileen Wade	Halina Whiffin
Liz Tavner	Doris Walters	Imogen White
Angela Taylor	John Walters	Sylvia Wilson
Roy Thomas	Richard Ware	Tony Wood
Sue Tiller	Peter Warwick	John Wright

Appendix 2

Examples of tracing study reports produced by the volunteer researchers

The volunteers were sent details by the researchers of the home they had agreed to investigate. These details were obtained from the Peter Townsend Collection archived in the UKDA. The relevant details relating to the home in 1958/9, where available, were transcribed by the research team onto a standard form that included the name of the home, its location, who owned it and who managed it, the number of residents it accommodated and any other information of potential use in tracing its history.

Each volunteer was provided with a briefing pack which included two sample reports together with suggestions about how to conduct their investigations and potential sources of information. Below are four examples of the kinds of reports we received on an ex-PAI, a former 'other' local authority home, a former voluntary home and a former private home. Many reports we received, including some of those reproduced here, were accompanied by substantial amounts of supplementary material including books and booklets, photographs, newspaper cuttings, maps, copies of official records and other supplementary material.

In the reports reproduced below, some details have been edited out to preserve privacy and confidentiality.

H 31 (An ex-PAI)

1. Name of home and address
(Please include change of name SINCE 1959
if applicable)

Tower Hospital, 46 Cambridge Road, Ely, Cambridgeshire.

2. When did the home cease to function as a residential care home for older people?
(Please supply details, if possible, of what happened to any existing residents at this point)

It is interesting that in 1990 there were 88 names on the electoral register of the residents and staff.

In 1991 there were none. Possibly the inmates were no longer considered permanent residents.

3. Why did it no longer function as a residential care home?

In 1988 the Tower buildings were considered to be substandard and beyond economic repair. In the next five years there were a number of attempts to find a private partner to build a facility with long stay beds for the elderly and a nursing home attached which would be contracted to the NHS. All bidders withdrew. When in 1992 the RAF decided to close their hospital in Lynn Road a deal was done whereby the Cambridge Health Authority agreed to pay £2.5 million to buy the hospital and it was decided to provide a district hospital for the town combined with the care facilities of the Tower with accommodation for the elderly patients. £500,000 for the purchase of the RAF Princess of Wales hospital was to come from the sale of the Tower site for redevelopment (Cambridge Evening News, 19 February 1992).

In March 1993, a US taskforce of 60 men, 30 volunteers and hospital staff moved the health centre at the Tower and all the physiotherapy and day hospital equipment from the Tower to Lynn Road. On Saturday 15 May 1993 a team of US Airmen from RAF Mildenhall emptied the old Victorian Hospital in a fleet of 40-foot trailers. Beds, furniture and office equipment were moved to the refurbished Princess of Wales Hospital in Lynn Road, Ely and the 62 elderly inmates were taken to their new home. The Tower was officially decommissioned on that date and together with the health centre was put on the market (CEN, 13 May 1993).

There were 66 beds at the refurbished Princess of Wales Hospital for elderly people needing long-term care. The centre was run by Lifespan Healthcare Trust with outpatient and other facilities specialising in geriatric medicine and care, as in the health centre at the Tower.

4. Was it demolished? If so, when?
(Please give reason if known)

Although put on the market in 1993, in 1998 the listed buildings of the old hospital were in considerable disrepair due to neglect and vandalism, but it was still considered 'an important city building' and it was hoped to preserve it.

5. What, if anything, was subsequently built on the site?
(Please supply information, with dates, on the use that the 'new' building/s was put to and whether its/their use has changed over time)

On 15 January 1998, planning permission was given to Springboard Housing Association to build 25 one-bed, 38 two-bed and two three-bedroom homes. The former chapel and staff housing were also included in the plans. A condition was that the main original building should be preserved as much as possible (Ely Standard, 15 January 1998).

While much has been demolished, many old features of the Tower buildings remain in the terrace of houses set around a court with communal gardens, called Tower Court, including the original tower itself, now converted into flats. The houses are desirable and resell at reasonably high prices.

6. If it was not demolished, what has the building been used for since 1959 and who owns it currently?
(Please supply dates if possible, including periods when the building was not used for any particular purpose)

See above.

7. Briefly describe the site/building and its location. Have there been any major changes to the building and/or to the local neighbourhood since 1959?
(Please supply details if possible, for example, of adaptations or extensions to building, major redevelopment of area and/or changes in demographic profile of neighbourhood)

The Ely Union workhouse was built in 1836–7 to a design by William Donthorn. It was built to accommodate 300 inmates and cost £7000. The building contractor was John Sugars of King's Lynn. The workhouse, built in yellow brick, was a variation on the standard cruciform plan, having much foreshortened front yards, and larger diagonally divided yards to the rear. At the front was the impressive three-storey Tower used as the administration block (7). Its fine main room was used for meetings of the Guardians and the Ely Rural District Council (3: Kelly 1933). There was an old women's yard on its north side, and the old men's yard to its south. To the west of each yard, respectively, were the two-storey old women's and old men's accommodation blocks. The Ordnance Survey 1901 gives a detailed picture of the site. On the site was a chapel and Kelly (1933) describes it as capable of holding 165 inmates and 95 casuals.

In 1917, the Isolation Hospital previously in St Johns Road moved into the grounds of the workhouse and the two institutions merged (3: Kelly 1937) and a row of huts was built for patients with contagious diseases. A new building was later built, funded by a charitable donation.

The last Public Assistance Minute Book (1) of the workhouse for the end of World War II up to the takeover by the National Health Service showed how difficult it was to run, with antiquated plumbing and very little modernisation. After the takeover, the number of workhouse inmates (i.e., very poor and homeless) declined. The Tower then appears to have become a care home for the long-term ill. Although

refurbished with 'light airy rooms', personal furniture, carpets and covered corridors linking the
buildings (4. Denton 1986), it is not surprising
that by 1988 it was reported as considerably substandard, with large, cramped rooms, no privacy,
and very difficult for staff (5: CEN 1988). The
lifts were inadequate, the electrical wiring was
faulty, the sanitation needed improvement, and there
was inadequate storage space. It was considered
a fire hazard. £100,000 was spent on alleviation,
but it was clear that the building was beyond economic repair.

The City of Ely has expanded greatly in the last
15 years and the Tower site, once 'a quarter of a
mile out of the city' (3. Kelly 1921) is now a desirable area within the old city near the Cathedral
and town centre.

8. What, if any, changes of ownership have there been since 1959?
(Please supply details with dates if possible)

1993 put on market by the local authority. In 1996
purchased by the Springboard Housing Association (5:
CEN). Since completion of Tower Court, houses have
been sold and are now privately owned (7).

9. Would you like to add any further information?

It became clear that the original Victorian workhouse was an unsuitable building for a modern care
home, despite being much loved and supported by
the townspeople. From 1988 the local authority was
searching for finance for a new residential and
nursing home. The move to the Princess of Wales
hospital seemed providential at the time of the
move in 1993. Unfortunately the Health Authority
was about to begin the process of closing down
long-term care in their institutions.

In 1996, community care teams were created by
the local Health Authority in East Cambridgeshire

including nurses, health visitors, chiropodists and therapists attached to GP practices. Major medical care was to be undertaken at the main regional hospitals of Addenbrookes in Cambridge, Hinchinbrooke in Huntington, and the district hospital at Ely. The Princess of Wales was to provide rehabilitation services, outpatient clinics with visiting consultants, occupational therapy and some inpatient care. By 1997 there were only four respite beds remaining. It was expected that long-term care would be transferred to nursing homes. However, by November 1998 there was a crisis in the hospitals as 35 of the 66 rehabilitation beds in the three district hospitals in Chesterton, Ida Darwin and the Princess of Wales were blocked by 'patients awaiting alternative appropriate facilities elsewhere' (CEN 20/11/1998).

10. Please list the sources of information you have drawn on to answer these questions. (Include contact details of individuals approached)

1. My first search at the County Record Office was not very productive, although I was able to read the Minutes of the Public Assistance Committee up until 1948, with interesting details of the last days of the old system during the war, and the preparations for the changeover to the new welfare state.
2. At the Cambridgeshire Collection of the Central Library I searched the electoral registers, which gave exact addresses, and numbers up to 1990. After which when they ceased to include No 46. There were 151 registered in 1957, and 82 in 1988.
3. Kelly's directories 1921, 1933, 1937 provided descriptions of the workhouse in the nineteenth and twentieth centuries.
4. A booklet, *Ely Union Workhouse*, by Audrey Denton (1986) was helpful on the hospital buildings in the grounds and the post-war refurbishments.

5. I was having some difficulty in finding further information when fortunately the present librarian, Chris Jakes, and the original librarian Mike Petty, of the Cambridgeshire Collection came in together from a meeting. One remembered that the closure of the Tower was connected with the Princess of Wales Hospital, the other remembered that cuttings files had been made, which they were able to find for me. The cuttings from the Cambridge Evening News (CEN) and the weekly Ely Standard provided the source material for the closure and move.

6. The Internet provided a useful page on the workhouse using much material I had already found at http://www.workhouses.org.uk/index.html?Ely/ Ely.shtml photographs of the Tower Court development.

7. Further details of the houses for sale or rent can be found on estate agents sites.

Signed: Francesca Ashburner **Date:** 26 October 2006

H 89 (A former 'other' local authority home)

**1. Name of home and address
(Please include change of name SINCE 1959
if applicable)**

Fairhaven, 236 Tunstall Road, Knypersley, Stoke-on-Trent ST8 7AB.

Now: The Roaches School.

**2. When did the home cease to function as
a residential care home for older people?
(Please supply details, if possible, of
what happened to any existing residents at
this point)**

Fairhaven ceased to function as a residential care home for older people on 11 July 1992 (1). The exact number of residents before closure is not known. In October 1991, 14 people were registered on the

Electoral Roll as living at Fairhaven (2). It is probable that care staff would register at their permanent home address, and so it is likely that all those people recorded at Fairhaven would have been elderly residents. But an unknown number of residents may not have registered.

After proposals for Fairhaven's closure were announced to residents on 11 February 1992 (3), the Biddulph Chronicle of 21 February 1992 recorded much local concern about what would happen to the residents, despite the Chairman of the Social Services Committee's assurance that 'residents would be consulted and asked where they wanted to go' (4). The alternative homes were all some distance away in Stoke-on-Trent and were not easily accessible by public transport from Biddulph (4). However, minutes of the Social Services Performance Review Sub-Committee of 29 July 1992 reported that 'once residents in the homes for the elderly had accepted there was no alternative but to close they wanted to move quickly and consequently they were transferred to homes of their choice as soon as possible' (5). This statement is supported by the fact that Fairhaven's closure had originally been planned for October 1992 (6).

Neither a search of relevant Council minutes (7) nor a letter to the local Social Services Office (8) have revealed the actual destinations of Fairhaven's residents. Reports in the Biddulph Chronicle of local Councillors' statements about the lack of alternative places in Local Authority homes in Biddulph and available spaces being in the north of Stoke-on-Trent (4) would suggest that most Fairhaven residents would probably have transferred to other Local Authority residential care homes in the north of the City.

3. Why did it no longer function as a residential care home?

On 14 February 1992 Staffordshire County Council Social Services Committee announced the proposed closure of Fairhaven, along with four other residential care homes in the County (4). The Biddulph Chronicle of 21 February 1992 reported the Social Services Committee Chairman's statement that the County had to make £3,000,000 cuts in the budget. Proposals in the Government's Community Care plan to move people out of homes and into the community meant that closure of some elderly persons' homes had to be considered. It appeared that these particular five homes in Staffordshire were selected for closure because they would have required considerable expenditure to bring them up to date. According to the County Council report, development proposals to meet safety standards and new laws in these homes would have cost the County Council £18,000,000 (4). Fairhaven, a converted Victorian parsonage without a lift (4) would have been expensive to update, and the Biddulph Chronicle of 6 March reported that the closure of Fairhaven would save the County Council £80,000 in 1992-3 and £150,000 in 1993-4 (9).

Additionally, the Social Services Committee Chairman spoke of the need to develop child care, home care and services for severely disabled people and the associated costs, for example, £3,000,000 to implement the new Childrens Act (4).

So Fairhaven was closed because it would cost too much to modernise, because of the new emphasis on care in the community and because of the need to spend more of the Social Services budget on other services.

4. Was it demolished? If so, when?
(Please give reason if known)

NA

5. What, if anything, was subsequently built on the site?
(Please supply information, with dates, on the use that the 'new' building/s was put to and whether its/their use has changed over time)

NA

6. If it was not demolished, what has the building been used for since 1959 and who owns it currently? (Please supply dates if possible, including periods when the building was not used for any particular purpose)

1959-11 July 1992	Continuing use as County Council Residential Home for Elderly.
July 1992-1995	Unused.
1995-2000	See below.
2000-2006	In current use as a residential home and school for children of secondary school age.

Minutes of the Social Services Committee of 4 September 1992 (1) recommended the sale of Fairhaven and other properties as surplus to requirements, and on 19 October 1992 the Director of Property Services was authorised to dispose of Fairhaven (10). Fairhaven was subsequently listed as available for sale throughout 1993 and 1994 (11), and, on 21 June 1995, it was recorded in minutes among a list of sales completed (12). There was no mention in the minutes that the Home had been in use during the period it was for sale.

Subsequently the names of two people appear on the Electoral Register for 1996 and 1997 as living at Fairhaven (13), but not thereafter. A local resident told me that Fairhaven had remained empty until it was opened as a residential school/home some six years ago (14). However, as the sale of Fairhaven by the County Council appeared to be to a private person, and not to an organisation (15), it may be that the former home was used as some sort of private residence for up to two years from 1995 to 1997.

The Land Registry records the current owners of Fairhaven securing a mortgage on the property on 1 October 1997 (16). As they are the proprietors of the School (17), it is likely that from this date preparations were made to establish the School, although it is not known at what date the School opened (18).

However, by March 2000 Fairhaven was registered with the Department of Education and Skills as a school and residential home for adolescents and it is in use for that purpose today (17). It is now known as 'The Roaches School' which is part of an organisation called 'Care Today'.

7. Briefly describe the site/building and its location. Have there been any major changes to the building and/or to the local neighbourhood since 1959?

(Please supply details if possible, for example, of adaptations or extensions to building, major redevelopment of area and/or changes in demographic profile of neighbourhood)

Fairhaven was originally a Victorian parsonage. It was built, along with the church opposite and the adjacent Sunday Schools building, for James Bateman, the owner of the Knypersley estate, in 1848-50 (19).

Fairhaven is a Grade II Listed Building (20). It is an attractive, substantial building built of local stone, which would have provided an imposing home for a Victorian parson (21). It has mullioned windows, a slate roof, tall chimneys and some decorative stonework on the roof above the original main entrance to the house. It is surrounded by a large garden and trees and bordered on all sides by a stone wall. From the A527 road to its east it looks very much as it must have done in the Victorian era, though sports equipment and sculptures in the garden provide evidence of its current use. A relatively small kitchen/dining extension of 1960's appearance has been built near to the side entrance (now the main entrance) to the Home, which is accessed off a small side road.

This Victorian group of buildings (Fairhaven, the Sunday School and the Church) are a striking and attractive part of the townscape, situated just beside the crossroads at the centre of Knypersley, a village located on the busy main A527 road between the Potteries and Congleton. Knypersley is in a valley on the edge of the Staffordshire Peak District, about a mile south of its much larger neighbour, the town of Biddulph, which provides all the small-town functions and amenities for Knypersley and the surrounding area. At the time of Townsend's visit, many men would have been employed in local coal mines and steelworks, only a mile away (19). Steel fabrication continues, but the mines are now closed. Almost 30 per cent of the working population is now employed in manufacturing, with approximately 20 per cent employed in the retail and associated trades sector and 11 per cent in health and social work, teaching and other professional categories (22). 99.1 per cent of the local population is white (22).

In many respects Fairhaven's immediate neighbourhood would have been quite similar in appearance in 1959 to that of today. Larger Victorian and Edwardian houses are strung out along the main road opposite Fairhaven towards Biddulph. The garage nearby remains, though it is now devoted to car sales. To the east of the main road is the large housing estate built for the National Coal Board in the 1950s, and on the south side of Knypersley crossroads 1920-30's ribbon development of detached and semi-detached houses and some local shops spreads along the main road towards the well-known Knypersley Cricket Ground and the steelworks further beyond.

Although there is no new housing in the immediate vicinity of Fairhaven, there has been considerable housing development in the area surrounding it since 1959. Many new private housing developments were built in the 1960s and 1970s. These are adjacent to the 1950's NCB estate to the east, and to the west of Fairhaven down into the valley.

Views from Fairhaven have not only been altered by new housing developments since 1959, but also by the construction of the A527 Biddulph bypass and a new associated roundabout immediately to the north. However, the countryside remains constantly in sight, with views of open fields and hills, both towards the Peak District in the east and westwards towards the folly of Mow Cop.

8. What, if any, changes of ownership have there been since 1959?
(Please supply details with dates if possible)

1995 Sold by County Council to an individual.
1997 Current owners, secure mortgage on property. It is not known if either of these people bought the property from the County council in 1995 or bought it from another person between 1995 and 1997.

9. Would you like to add any further information?

1. I discovered that minutes of various Staffordshire County Council Social Services and Property Committees are not fully open for public consultation. However, I received comprehensively researched replies to all my questions in a letter from Rebecca Jackson, a County archivist. The only item missing is the name of the purchaser of Fairhaven, whose confidentiality is protected in line with Council policy. (See letter and research findings dated 17 May 2006.)
2. On the Home Details Form, the date of Fairhaven's construction is given as 1804. It was in fact built in the period 1848-50.

10. Please list the sources of information you have drawn on to answer these questions.
(Include contact details of individuals approached)

1. Staffordshire Record Office (SRO): Staffordshire County Council Social Services Committee (SCC

SSC) minutes, 4 September 1992 (CC/B/132/22). Contact Ms Rebecca Jackson, Archivist, Eastgate Street, Stafford, ST16 2LZ. Tel. 01785 278379.

2. Electoral Register, Staffordshire Moorlands, 1991. Contact Carol Wedgwood, Staffordshire Moorlands District Council, Moorlands House, Stockwell Street, Leek, ST13 6HQ. Tel. 01538 483405.

3. SRO: Staffordshire County Council Social Services, Kidsgrove and Biddulph Advisory Sub-Committee (SCC KBASC) minutes 1989–92, 14 April 1992 (CC/B/133/16/1).

4. 'The Biddulph Chronicle', 21 February 1992. Biddulph Library, Tunstall Road, Biddulph, Stoke-on-Trent, ST8 6HH. Tel. 01782 512103.

5. SRO: Staffordshire County Council Social Services Performance Review Sub-Committee (SCC SSPRSC) minutes, 29 July 1992. These are attached to SCC SSC minutes, 4 September 1992 (CC/B/132/22).

6. SRO: SCC SSC, 14 February 1992 (CC/B/132/21).

7. Research findings from Rebecca Jackson, Archivist, 17 May 2006, Item 2.

8. Letter to David Eckersley, Team Manager, Kidsgrove, Staffordshire Social Services, 29 March 2006.

9. The Biddulph Chronicle, 6 March 1992.

10. SRO: SCC Property Sub-Committee (SCC PSC) minutes, 19 October 1992 (CC/B/156/6).

11. SRO: SCC PSC minutes, 18/12/1992, 14/6/1993, 14/9/1993, 2/11/1993, 11/1/1994, 21/3/1994, 22/6/ 1994, 12/9/1994 (CC/B/156/6-9).

12. SRO: SCC PSC minutes, 21 June 1995 (CC/B/156/9).

13. Electoral Register, Staffordshire Moorlands District Council, 1996 and 1997. Laurel Bundred and Nicholas Bundred were registered as resident at 'Fairhaven'.

14. Personal conversation with owner of Car Sales Showroom near Fairhaven.

15. SRO: SCC PSC exempt minutes, 21 June 1995 as reported in research findings of Rebecca Jackson, archivist, 17 May 2006, Item 4.
16. Land Registry Title Document for Fairhaven, Title No. SF318419: Internet.
17. H. M. Chief Inspector of Schools, Ofsted Report (from Internet). School number is DfES No. 860/6017.
18. Letter to The Roaches School, 29 March 2006.
19. 'Biddulph, A Local History', Ed. Joseph Kennedy. 1980, Department of Adult Education, the University, Keele, Staffs, ST5 5BG.
20. Planning Department, Staffordshire Moorlands District Council, Moorlands House, Stockwell Street, Leek, ST13 6HQ. Tel. 01538 483576.
21. Photographs of Fairhaven, taken February 2006.
22. 2001 Census, Biddulph West Ward.

Signed: Angela J. Taylor   **Date**: 1 August 2006

H 125 (A former voluntary home)

1. Name of home and address
(Please include change of name SINCE 1959 if applicable)

1959: Grendon House, Grendon Road, Exeter (1).
2005: Gainsborough House, Grendon Road, Exeter EX1 2NH.
Flats 1-22 (2).

2. When did the home cease to function as a residential care home for older people?
(Please supply details, if possible, of what happened to any existing residents at this point)

Summer 1974 – It was noted in the minutes for 4.3.74 of Exeter Municipal Charities (EMC) that on 8.5.74 the last two remaining residents would be accommodated by Exeter City Council Social Services department. On 1.4.74 the EMC minutes noted that the matron would be in post until the end of June 1974 (3).

3. Why did it no longer function as a residential care home?

On 25.1.73 The Exeter Municipal Charities minutes relate that [the matron] was 'invited to resign', which she agreed to do. This followed a period of various staffing difficulties involving the matron (3).

The Home Committee met on 12.2.73 to review the future of the Home. 'The average number of occupants was six in recent years – although there were only four at present and the home can accommodate 12 officially but is capable of accommodating 15'. It was known that council homes had a waiting list and were building more homes. Three alternatives were considered: (a) to sell; (b) to lease to the council; (c) to admit persons on the council waiting list so it was full. These alternatives were to be explored with the social services department (3). On 14.3.73 when the Home Committee met with social services representatives they commented that the minimum economic number for a home is 38–40 so they could not take on Grendon House for use as residential care home (3). On 21.6.73 the Home Committee agreed to sell the Home privately and use the money to build Livery Dole Almshouses (3). This was on a site/garden of nearby Almshouses already owned by the charity (6).

On 15.11.73, outline planning permission for 22 units (3).

4. Was it demolished? If so, when?
(Please give reason if known)

In the minutes of 11.7.74 it was noted that Grendon House was sold to The Devon Community Housing

Society for £55,000 (3). The house was demolished in November 1974 (3).

5. What, if anything, was subsequently built on the site?
(Please supply information, with dates, on the use that the 'new' building/s was put to and whether its/their use has changed over time)

Gainsborough House was 'inaugurated on 28.10.77 (7). It comprises 22 one-bedroom flats. It is a sheltered housing scheme with a warden on site, a 24-hour alarm system using a call centre. The criteria for residents is that 'they are 55 years or more and in need of some sort of support' with daily living. This does not include being in need of care. A pre-tenancy assessment is undertaken by the housing officer. There have been no changes to the use of the building or the criteria for occupancy since 1974 (5).

6. If it was not demolished, what has the building been used for since 1959 and who owns it currently?
(Please supply dates if possible, including periods when the building was not used for any particular purpose)

NA

7. Briefly describe the site/building and its location. Have there been any major changes to the building and/or to the local neighbourhood since 1959?
(Please supply details if possible, for example, of adaptations or extensions to building, major redevelopment of area and/or changes in demographic profile of neighbourhood)

Gainsborough House is a purpose-built two-storey block of flats with gardens for the use of the residents. It is an 'open site', each resident has their own dependent entrance and landings; there is no communal lounge area for residents. Grendon Road is in a residential suburb of Exeter about three quarters of a mile from the city centre and adjoins one of the busy main roads to the city centre. Nearby are local facilities and shops in Heavitree and St. Leonard's (6). It is a short road with two blocks of Almshouses adjacent to Gainsborough House, both still belong to the Exeter Municipal Charities. They were built in 1880 and 1892 with 24 self-contained flats. On the other side of the road are two large blocks of flats built in the 1970s, one is on the site of an old house, Grendon Lodge, which was demolished in 1970s. One block consists of 69 units and the other 48 units. They are owned by a Housing Association and Exeter City Council (7).

The surrounding area has not changed significantly with a mixture of family homes, private rented accommodation with many moderately sized houses dating from the 1880s to the 1930s with gardens (7, 6).

8. What, if any, changes of ownership have there been since 1959?
(Please supply details with dates if possible)
1959 Exeter Municipal Charities.
November 1974, Devon Community Housing Association (3).

9. Would you like to add any further information?
Attached are details of the Exeter Municipal Charities and the email contact with its correspondent.

10. Please list the sources of information you have drawn on to answer these questions.
(Include contact details of individuals approached)

1. Kelly's Directory for Exeter 1960
2. Bray's Exeter Street Directory 1987/88
3. Exeter Municipal Charities Minutes – January 73–December 76
4. Exeter Municipal Charities Minutes – January 63–December 65??

5. Mr Crook, Housing Officer, Devon Community Housing Association 01392 686438
6. Investigator's personal knowledge
7. Site visit

Signed: Julia Spruntulis   **Date:** 8 December 2005

H 139 (A former private home)

1. Name of home and address
(Please include change of name SINCE 1959 if applicable)
Roslyn, [...] Grosvenor Road, Margate, Kent.

2. When did the home cease to function as a residential care home for older people?
(Please supply details, if possible, of what happened to any existing residents at this point)
It would appear that Roslyn ceased to function as an old people's home between 1967 and 1969.

3. Why did it no longer function as a residential care home?
As noted in section 6, this house reverted to holiday accommodation but I have not been able to determine a reason for the change. Also see section 10.

4. Was it demolished? If so, when?
(Please give reason if known)
The building is still extant.

5. What, if anything, was subsequently built on the site?
(Please supply information, with dates, on the use that the 'new' building/s was put to and whether its/their use has changed over time)
NA

6. If it was not demolished, what has the building been used for since 1959 and who owns it currently?

(Please supply dates if possible, including periods when the building was not used for any particular purpose)

By 1969, still using the name Roslyn, Mrs D Thorne was operating as a guest house. In 1971 it was still listed as a guest house but no proprietor is listed. By 1972 Michael Mazza is listed as the owner or occupier. The house is currently subdivided into flats and no value was attached to attempting to record the occupants after the above.

7. Briefly describe the site/building and its location. Have there been any major changes to the building and/or to the local neighbourhood since 1959?

(Please supply details if possible, for example, of adaptations or extensions to building, major redevelopment of area and/or changes in demographic profile of neighbourhood)

Built after 1874, Roslyn is a characteristic four-storey Victorian terrace house probably built speculatively for letting to families of middling income. Well situated on an elevated site, this terrace originally had near uninterrupted views to the West and the sea over largely unoccupied land and Margate's two newish railway stations. By the 1920s a good number of houses in the street were given over to holiday accommodation and the whole district built over, with Dreamland occupying the near foreground. The whole of central Margate has declined in status since the 1960s reflecting the collapse of the holiday trade and the general shabby appearance has only recently been arrested.

8. What, if any, changes of ownership have there been since 1959?

(Please supply details with dates if possible)

See Section 6.

9. Would you like to add any further information?

There were at least eight guest houses […] [in Grosvenor Road] in 1959 but no other house in this street operated, openly at least, as an old people's home. Even Roslyn was not entered in Kelly's as such for its early years. I note from the initial details provided by the research team that the home was 'run by Mrs Thorne, and Mr Thorne's daughter' rather implying that Mr Thorne had employment elsewhere or was incapacitated and lived at home. If the latter, then there is the possibility that this was a reason for the change of use. I imagine the Archive will shed some light on this. It may be relevant to note that Margate as a seaside resort had been in steady decline since 1918 and the change of use by the Thornes might well have been an astute move to secure a steadier all-year-round income.

10. Please list the sources of information you have drawn on to answer these questions. (Include contact details of individuals approached)

I made a site visit on 12 January 2006 and was pleased to find immediately an elderly resident who was categorical that […] [Roslyn] was not an old people's home when she had moved into a near adjacent property in 1972. At Margate library I found a good run of Kelly's (for the most part in alternate years) which provided an adequate chronology for Roslyn.

```
1948   No entry for [No. …] but adjacent
       properties were guest houses
1951   Roslyn Mr & Mrs Thorne
1955   Roslyn Mr & Mrs Thorne
1957   Roslyn Mr & Mrs Thorne
1959   Roslyn Guest House Mr & Mrs Thorne
1961   Roslyn Boarding House
1963   Roslyn Old People's Home W Thorne
1965   Roslyn Old People's Home Mrs D Thorne
```

1967   Roslyn Old Peoples Home Mrs D Thorne
1969   Roslyn Guest House
1971   Roslyn Guest House No name entered
1972   Michael Mazza

The closure of Roslyn well before regulation means it does not feature in any deposited records. This I confirmed with the Care Standards Commission Ashford and KCC Social Services Department, Maidstone. I also approached the Kent Care Homes Association regarding any members' recollections of Roslyn but to date have had no response.

Signed: Terry Powell   **Date**: 27 January 2006

Appendix 3

In this appendix we present extracts from four diaries: two male and two female residents, two written in 1958–9 and the other two in 2006.

A day in a 1959 diary

7.30   Attendants came round and called us. Got up and took my turn at the washbasin at the end of the passage, then I dressed and went downstairs. Mary (sleeps in the next bed) wasn't feeling too good and the Sister said she could stay in bed.

8.15   Had breakfast (tea, porridge, bacon, bread and marmalade). Went to kitchen and helped with washing up for half an hour. I get five shillings a week for helping. Cook hurried us out because she wanted to get on with her work.

9.30   Went to No. 4 day room and had a look at the 'Mirror'. Talked to my friend Milly about her daughter. Did a bit of knitting. Mrs Jones had a row with Mrs Robbins. Mrs Robbins wanted the window open. (she's a bit of a trouble maker). In the end the attendant came in and they quieted down.

11.00   Went out for a little walk on my own and had a look round Woolworth's. Bought some hairpins.

12.30   Went back for dinner (Roast lamb, cabbage and potatoes, stewed apple, cup of tea).

1.30   Went to No. 4 day room. Read book for an hour but my eyes got sore. Mrs Brown told me her rheumatism is playing up

again and said she would tell Matron
when she came round.

3.00   Bath day. Waited 15 minutes in changing
room and then the attendant bathed me.
I used to bath myself at home but we are
all helped here. I suppose some of them
need it. Some old people might hurt
themselves on their own.

5.00   Tea (herring, bread and butter, jam, tea).

6.00   My Nellie (youngest daughter) dropped
in and we talked for half an hour. She
told me about the new self service shop
down in the High Street. She brought
me 10 cigarettes and told me that her
boy (he's 12) has got a bad cold.

7.00   Watched TV for an hour. Didn't want any
supper (milk and biscuits). Had a smoke.
I'm beginning to get used to living
here now. I was lonely at first and
everything was strange. People come here
from miles around and there's only one
from my home district. They're a funny
lot but I keep my sense of humour.

9.00   Went to bed. I was very tired and had a
bit of a headache. Mary was feeling better
and will be getting up tomorrow. She was a
bit fed up because some of us came in
just as she was nodding off.

A day in a 2006 diary

Midnight to 7.00 Sleeping. I am on 4th floor.

7.00   In bed waiting breakfast. Christine
came (corn flakes, tea, bread butter &
marmalade).

8.00   Got up and washed.

9.00   Got dressed. Jodie came to make bed.
We had light conversation. It is dull

today but I am expecting 2 visitors,
my neighbour when I lived in Norbury.
We got to be friends and she visits me
pretty often about every 4 weeks.

10.00 Carer brought me a glass of milk
(not keen on tea & coffee).

11.00 I have lost my balance so can walk
around the room with zimmer but taken
to meals wheelchair. I have a lovely
room very sunny and a lovely view.

12.00 Had dinner. Relish. 3 Veg Mash Pots Sweet
Jelly Mandarins Ice Cream. Resting on bed.

1.00 Resting.

2.20 Visitors arrived, stayed until 4.15pm.
Very nice afternoon.

4.00 Get ready for tea. Most had fish finger &
chips not me. Asparagus. Have trouble
with swallowing.

5.00 Back in my room. Richard (from India) very
nice carer.

6.00 TV News.

7.00 Start doing bits & pieces ready for night.
A friend from Bournemouth phoned me. My
grandson's wife telephoned me for a chat.

8.00 Into bed, watch TV. My grandaughter
telephoned me to chat and said she with
her friend would come and visit me Sat
morning. She is a lovely girl. I don't see
her a lot but she is a lovely character.

9.00 Radio as TV is horrible after 9 pm.

10.00 Take tablet. Sleeping.

A day in a 1958 diary

I am awake early on the last day of January
but in the night I could scarcely sleep with
one thing and another. I dream of my comming

76 Birthday. I am Old, they say, that is of the years. I have lived and toild. I have worked hard all my life, and I must say One Benefits by hard work because it occupys ones mind. Makes you sleep, tired and keeps ones musels and limbs Active by using Energy. At 6-30 I wash myself, clean my bed Pot out, make and tidy up my Bed. Then I go down to my usual cup of Tea. Here in the Dining Room assemble a few of the men, Inmates, who all have a cup of Tea. Now, myself get my clean coat on and apron and help the Staff to get tables ready for Breakfast. Clean Cups, Saucers, Plates, Knives and Forkes, Spoons. This I help to do. Then Officers order, to down the Kitchen for food to be ready shortly. Of course the food varies every day which is clean and wholesome. Being that I am considered to be very alert and quick in all I do. The Warden was very considerate to me and appreciated my Work. He also was kind to me. [...]

I felt so miserable, because I am very active and always, occupied my idle time in doing something useful, the same, when I am at My Home, should I call it my Home, which was in Nottingham. Married to My second Wife who has turned me Out. This has caused Me to be here. I applied to the 'Warden' for a Job, something to do, occupy my mind. I told him I could do almost anything, no matter what. First it was sugested I should work on a Stitching Machin, attaching some leather binding on some frayed carpets, because I had told him, I once had a Boot and Shoe Making and repair Shop in Salford, Lancashire, also I had a Paint and Decorators Buisiness, as I was a Painter, and Own Master. It was in 1919, after the 1914 War. I was asked to make a pair of Boots, for a prominent Town Hall Civil Servant Officer.

Now the Warden, and the Principle Orderly, sugest I might wish to go and help in the Dineing Hall.

Soon the last day of January I begin My Task
getting up at 5-30, to 6 Oclock. I go down to
the dineing hall, and arrange all Tables, Cups
Saucers, for Our early Cupa. Cup of Tea, which is
greatly Appreciated by All Men and Women Inmates.
But let Me remind You, That all Men, do Not get
up for this and desire, to have more rest in Bed.
They then get up at Severn Oclock. All Inmates
have to be up. Empty Slopps, make his or her Bed.
When this is done, all Men should go down, to the
Court Yard where the Wash House. When all have
washed, Each Man's time is his own.

As a rule, most go in the Day Room, to have a
Smoke. Cigaretts or Pipe, some will go out to the
local paper Shop, for a Mornings News, those who
do so generaly get other Infirm Mens Newspaper
and perhaps receive a copper, for doing so. I
must say here that every man gets Pocket Money
of 10 Ten Shillings a week, on Top of This Every
Man gets His Cigaretts of Ten, or Tobacco. If he
gets Tobacco, he is also given Cigarett Papers.
Personaly I think that is very Good for the
Powers to be, You doing this very Kind Action.
Even a Smoke helps Ones Nerves and Consoles his
Mind, whilst waiting for Breakfast, some Men have
a walk round in Court Yard, or go in the local
Street. Come back in time for Breakfast, which
is served at Eight O-clock Now all Assemble in
Dineing Room, some times there is a Rush for it,
'Why 'I personaly dont know Unless it is to get
to the table first and, then, to grab a piece
of extra Sugar there in the Saucer provided.
Each Man is alowed four peices of lump Sugar. Of
course perhaps, there a few short. But if The
Officers, find out, Well He will be reprimanded
an Questioned. All Tables are Sett by Me, with
the help of One or Two Officers. Cups, Saucers,
Knives, Spoons. Etc. Etc depending on what is
for Breakfast. If it is Porridge as it is every
morning Spoons are provided. If it is Fish all

have a Fish Knife specialy for that, also they
are provided with an Ordinary knife because,
some, men Cut the Outside Crusts off, a few Men
wrap these crusts up in a small paper bag, to
throw down in some bye way for the Pigeons. There
are many arround here, of course it is very kind
of the few men, who do this. But the Warden and
Staff dont approve, of this as they the birds
cause a deal of dirt arround the Vicinity. One
Man has the task to clean and Sweep up the Court
Yard. There are a few Men, who are able to do
various tasks. Sweeping. Cleaning Wash House and
Wash Basins. Clean out the Out Side W.C. Toilets.
There is the Two Day Rooms to be kept Clean.
All This is done directly after Breakfast. If
the weather is bad or is very cold, The Two Men
appointed, to clean out the Fires and remake, for
warmth. There is also a Heating Radiator which is
always on all Night in One Day Room. Now untill
the fires are made a few Men, get close to this
for warmth. Of course, One can quite Understand
this, as some mens Blood feels cold, on early
Morning. They warm their hands on this. Some
will put a few handercheifs, on when washed, to
dry and they soon dry here on it. As I stated
previously, The Meals Vary. Some mornigs it is
Porride, Marmalade, a spoonfull and some use it
to sweeten thiere porridge. Tea is provided and
some Men buy Sugar to add to the Tea. Fish and
Tomatoes. Brown and White Bread is provided and
Plenty if asked for. No One Need go Hungry and
The Staff are Very Obliging and Kind, in All They
do. Of course They are very Strict. One has to be
in a place like this. My personal oppinion, This
is a Home from Home, for some Men Some have lost
theire All, Their Wives, Sons Daughters Etc and
One Never Knows Who they may have been in Theire
previous days. There are a few Men here Who are
Very Intelegent, and in the Know ... [*This is
the end of the fourth page, the next eight are
missing*]

... cough up Phlem. I notice that One or two Spit
this up into theire handkerchiefs, or in a small
tin or bottle. The handkerchief, perhaps in their
pockets for days on End till they feel inclined
to throw away or wash it. I don't know but its
very Unigenic, unhealthy for others.
I am still writing these words here in the large
dinning room. It is Now Twenty to Ten O clock by
our Dinning hall Clock. I am very tired after a
hard days busy work and [removening?] I now stop
writing for to night and go into the small
dayroom, to bid Good Night to All and God Bless.
When I got upstairs to my Bed Room I spoke to
some of the Men next room and bid them, Good
Night and God Bless to All particularly to Dick
partly blind and to a Yorkshire Man, a Yorkshire
Man, like myself, as I was passing the down
stairs day room, I overheard One particular Man,
telling one of the Officers untruths. Lies. He
told the Officer I was shouting and creating a
Nuisence. I spoke softly as some Men were asleep,
as I was very tired, I soon went to sleep after
I said my prayers.

A day in a 2006 diary

6.00 Midnight – 3.00am doing Sudoku puzzles &
     some cigarettes. To bed 3.00am.

7.00 Asleep.

8.00 Ditto

9.00 Ditto

10.00 Ditto

11.00 Ditto

12.00 Listening to radio in bed.

1.00 Ditto

2.00 Had haircut (£4). Then collected lunch
     from main kitchen: beef ... Ate lunch
     in own room.

3.00   To smoking room 2.15pm. Made cup of tea.

4.00   Bought Daily Mail (40p) from Shell
       garage. Made cup of tea 3.45 pm & smoked
       in smoking room.

5.00   To tea 4.35pm. One sausage roll & piece
       of cake. Poor teatime I thought.

6.00   Read aforementioned newspaper ('Grey day
       for Blair') & smoked in smoking room.

7.00   In smoking room with Sudoku puzzles
       in Daily Mail, its Scrabble-Grams,
       Mini-Sudoku & Sudoku X. Supper trolley,
       sandwiches.

8.00   Ditto

9.00   Ditto

10.00 Ditto. Snipped out Spot the Ball solution
       from my Daily Mail.

11.00 Watched part of Question Time on BBC TV.

12.00 Made cup of tea in smoking room. Forgot 9
       tablets daily medication until about
       midnight. Continuation of rather
       inadequate portion(s) of food at
       tea-time.

Appendix 4

The 2005–6 quality measure

We based a measure of the quality of care provided in each surviving home in 2005–6 on CSCI Inspection Reports. At that time, inspectors assessed the performance of homes they visited in England against each of 38 standards using the following scores:

Standard exceeded (commendable)	score 4
Standard met (no shortfalls)	score 3
Standard almost met (minor shortfalls)	score 2
Standard not met (major shortfalls)	score 1

Homes were inspected on average twice a year, although there was some variation depending on the outcome of previous inspections. Not all standards were assessed at each inspection. In producing an overall score for the homes we visited, we first selected the inspection report which was closest to the time of our visit and recorded all the scores available. We then worked backwards through the reports (generally speaking these were produced at approximately six-monthly intervals) until we had obtained the most recent score for each of the standards. For the surviving homes which were not visited, we started with the report nearest to December 2006 (when the fieldwork ended) and followed the same procedure.

Standard 6, Intermediate Care, was not included because most homes did not provide such care. As a result a home that scores 4 on all the other 37 standards achieves the maximum possible score of 148.

The following worked example lists the 37 standards and demonstrates how we calculated the quality measure. Scores were entered on the dates each particular standard was assessed by the CSCI Inspector.

Date of visit: 22/23 November 2006

Standards	Aug 2004	Dec 2004	July 2005	Feb 2006	Feb 2007
Choice of home					
1. Information to make an informed choice				3	
2. Written contract				3	
3. Full assessment					3

(*Continued*)

Standards	Aug 2004	Dec 2004	July 2005	Feb 2006	Feb 2007
4. Capacity to meet needs			3		
5. Trial period				3	
6. Intermediate care				N/A	
Health and personal care					
7. Individual care plans					3
8. Promotion of health and access to health services				3	
9. Medicine administration				3	
10. Respect for privacy					3
11. Dignity in death				3	
Daily life and social activities					
12. Flexible routines				3	
13. Visitors and local links				3	
14. Personal choice and autonomy					3
15. Balanced diet				3	
Complaints and protection					
16. Responsiveness to complaints				3	
17. Protection of legal rights				3	
18. Protection from abuse				2	
19. Accessible, safe and well-maintained home				3	
20. Communal facilities				2	
21. Toilet and washing facilities				3	
22. Specialist equipment available					3
23. Adequate space for each resident				2	
24. Comfortable and private bedrooms		3			
25. Satisfactory and safe surroundings				3	
26. Clean, pleasant and hygienic home				3	
27. Adequate staffing level				3	
28. Adequately trained staff					3
29. Thorough recruitment procedure					2
30. Staff training and development programme				3	
Management and administration					
31. Competent registered manager				3	

Standards	Aug 2004	Dec 2004	July 2005	Feb 2006	Feb 2007
32. Management style			3		
33. Effective quality assurance				2	
34. Suitable accounting and financial procedures				3	
35. Residents' financial interests safeguarded					3
36. Implementation of employment policies				3	
37. Good record-keeping procedures				3	
38. Promotion of health, safety and welfare					2

	Score	Number	Total score
Standard exceeded	4	0	0
Standard met	3	31	93
Standard almost met	2	6	12
Standard not met	1	0	0
Total score			105

Notes

1 Why Revisit *The Last Refuge*?

1. The National Assistance Act 1948 imposed a duty on local authorities to provide residential accommodation 'for persons who by reason of age, infirmity or any other circumstances are in need of care and attention' (s.21). Voluntary sector and privately run residential homes had to be registered with, and inspected by, local authority welfare departments. Historically, residential homes catered for less dependent older people than nursing homes. In 1948 nursing homes were registered with, and inspected by, local authority health departments under the Public Health Act 1936. In 2002, residential care homes and nursing homes in England and Wales ceased to be registered and inspected by separate bodies and the umbrella term 'care home' was adopted to include homes providing residential care and/or nursing care.

2. We would like to thank Sonia Jackson, the author, for bringing this review to our attention.

3. In 1963, Townsend became founding Professor of Sociology at the University of Essex. He moved to the University of Bristol in 1982 where he later became Emeritus Professor of Social Policy and returned to the LSE in 1998 where he was appointed Centennial Professor of International Social Policy. He retired from this position in 2008. Apart from *The Family Life of Old People* (1957) and *The Last Refuge* (1962), his many hundred publications include *Old People in Three Industrial Societies* (1968) which he co-authored with Ethel Shanas and others, *The Social Minority* (1973), *Poverty in the United Kingdom* (1979) and *The International Analysis of Poverty* (1993).

4. Under the Local Government Act 1929 the responsibility for the relief of the poor was transferred from the Poor Law Unions to the Public Assistance Committees of county boroughs and counties in England and Wales. The old workhouses, hospitals and other forms of indoor relief were included in this transfer. The workhouses were renamed 'public assistance institutions'. Under the National Assistance Act 1948 the former poor law system was replaced and residential services became the responsibility of the welfare committee of local authorities, but the inherited buildings did not finally disappear until the 1970s. They became known as 'former public assistance institutions'. In this book we use the abbreviation 'ex-PAIs'.

5. In 1958–9, services for older people were the responsibility of local authority welfare departments which were set up under the National Assistance Act 1948. The chief welfare officer was in charge of the department.

6. A guinea was 21 shillings (which in UK sterling in 2009 would be £1.05p). Three-and-a-half to 30 guineas would have been £3.13s.6d to £31.10s.0d.

7. At the time of writing this book, Robert Pinker was Professor Emeritus of Social Policy at the LSE. Sheena Rolph and Julia Johnson interviewed him on 23 February 2007.

8. The first section of the M1 motorway was not opened until November 1959.

2 Changing Contexts of Care

1. Comparable statistics are not available for the years prior to 1960. However, Townsend reported that on 1 January 1960 there were 3335 residential care homes in England and Wales, providing 110,767 places. Two-thirds of the 95,500 residents of pensionable age were in local authority premises (half of them in former workhouses), a quarter were in homes run by the voluntary sector and a tenth in premises provided in the private sector.

2. Laing & Buisson (2009) reported an increase in the number of care home places for the first time since 1995.

3. Census data for the UK for 1971 and 2001 show that the population aged 85 and over increased from 472,000 to 1,128,000. Over the same period, the total number of places in long stay facilities (including hospitals) increased from 267,800 to 530,100. This represents a decline from 57 per 100 head of population aged 85 and over to 47.

4. The Government of Wales Act 1998 created the first National Assembly for Wales. Unlike the devolution arrangements in Scotland (Scotland Act 1998) and Northern Ireland (Northern Ireland Act 1998), this Act did not separate the legislature from the executive. So although the Assembly had a large number of order-making powers it had no powers to make primary legislation. The Government of Wales Act 2006 dismantled these constitutional arrangements and created three entirely separate bodies: an executive body, the Welsh Assembly Government; an unincorporated association of 60 members, the National Assembly for Wales; and a support body, the National Assembly Commission. In consequence, health and social care policies for Wales have been diverging from those in England and other parts of the UK. For example, at the time of writing it was the only UK nation to have a Commissioner for Older People.

5. The newly elected Conservative government in 1979 believed that the mechanisms of the market were the most effective and efficient way of providing welfare services and creating choice. Through the Housing Act 1980, it brought in the 'right to buy' policy which allowed and provided incentives for certain tenants to buy their, previously rented, properties from the council. 'Special' housing occupied by older people was exempt, creating a residual category of housing (Peace and Johnson, 1998). The Housing Act 1985 allowed local authorities to dispose of blocks of housing stock with or without tenants and, through the Housing Act 1988, secure council tenants were given the right to transfer to a different landlord, such as a housing association.

6. Between 1980 and 1983, the DHSS had discretionary powers, through the Supplementary Benefits (Requirements) Regulations 1980, to pay means-tested board and lodging allowances to residents of private and voluntary homes. In 1983, the regulations were amended so that what had been a discretionary payment became a right.

7. Prior to this new legislation, the Nursing Homes Act 1975 required health authorities to register and inspect nursing and mental nursing homes and the Residential Homes Act 1980 required local authorities to register and inspect private and voluntary residential homes accommodating four or more residents. The latter was amended in 1991 to include smaller homes.

Under the Health and Social Services and Social Security Adjudications Act 1983, residential homes providing nursing care could, for the first time, be dual registered with both health and local authorities.

8. The Burgner Report (1996), for example, argued for a complete overhaul of these services and recommended that regulation should become the responsibility of local authority trading standards departments.

9. The Royal Commission argued that people in long-term care incur three kinds of costs: (1) living costs (food, clothing, heating amenities and so on), (2) housing costs (the equivalent of rent, mortgage payments and council tax), (3) personal care costs (the additional cost of being looked after arising from frailty or disability) (Sutherland, 1999a, p. 64). Its definition of personal care included all direct care related to the following:

- Personal toilet such as washing, bathing, skin care, personal presentation, dressing and undressing,
- eating and drinking, as opposed to obtaining and preparing food and drink,
- managing urinary and bowel functions, including maintaining continence and managing incontinence,
- managing problems associated with immobility,
- managing prescribed treatments, for example the administration and monitoring of medication,
- managing behaviour and ensuring personal safety, for example, for people with cognitive impairment.

Its definition also included the support and assistance which skilled professionals offer to enable people to do things for themselves, for example, offering reminders to people with dementia (Sutherland, 1999a, p. 68).

10. At the beginning of 2005 the King's Fund set up a group to examine the kind of social care that should be provided for older people over the following 20 years. The group was chaired by Sir Derek Wanless, the former banker, who had been responsible for a Treasury funded review in 2002 of the future of health spending (Wanless, 2002). The social care review (Wanless, 2006) was published shortly after the White Paper, *Our Health, Our Care, Our Say*, in March 2006.

11. In identifying alternatives for the future funding of long-term care, the Green Paper *Shaping the Future of Care Together* (DH, 2009) proposed several options. Two were rejected: no state contribution (on the grounds that many people would not be able to afford care) and full state funding like the National Health Service (on the grounds that it would place too high a burden on the taxpayer). The remaining options all provide to those deemed in need of care a basic minimum entitlement funded through taxation, dubbed a National Care Service. This entitlement would be combined with payments made by individuals, a version of the partnership model. In respect of the contributions by individuals, two options were put forward. One involved the state working with private insurance companies whereby individuals could volunteer to pay for their care costs, if it turned out that they needed care. The other option would involve all those at or above retirement age

(presumably meaning the age at which the person becomes eligible for a state pension) paying a means-tested amount into a compulsory state insurance scheme in return for any care needed being fully funded. In both options, the individual contributions could be paid as a lump sum, before or after retirement or via their estate after death. None of the options covered accommodation costs in care homes, as such costs apply to everybody irrespective of their care needs. To help cover these costs, the Green Paper suggested the establishment of a 'universal deferred payment' scheme, so that these costs can be paid for after death.

3   The Study Design and Methods

1. Townsend oversampled homes with more than 100 residents (mainly the ex-PAIs) because, although they only represented four per cent of homes in England and Wales, they accommodated 25 per cent of all residents.
2. See Chapter 1 note 4.
3. The acronym JU referred to 'Joint Units'. Many ex-PAIs were run in part as hospitals and in part as homes for needy or vulnerable people. Each one was to 'be treated as a single unit, the future ownership and management being determined by its predominant user' (Ministry of Health, 1947). However, on the appointed day (5 July 1948) some of the residential buildings were continuing to house sick people and some hospitals were still housing those in need of care and attention. 'Such accommodation became known as joint user institutions, and they were seen as a temporary expedient' (Means and Smith, 1998, p. 140).
4. All the surviving homes named in this book have been given pseudonyms to protect their identity. The original names of the non-surviving homes have been retained, however.
5. See note 4 above.
6. See note 3 above.
7. A Registered Homes Tribunal was the tribunal that heard appeals in relation to the registration of nursing homes, care homes and children's homes prior to the implementation of the Care Standards Act 2000.

4   Survivors and Non-Survivors

1. In June 2006 Southern Cross floated on the London Stock Exchange, with a listing valuing the company at approximately £550 million and it was reported that the company operated 578 homes with expected revenues in excess of £600 million (*Sunday Times* Business Section, 4 June 2006). The good times did not last suggesting yet further instability and change. At the end of June 2008 the company failed to meet a £46 million repayment deadline and was forced into emergency talks with its banks. The collapse in the property market meant it was not able to sell assets to repay costly bridging funds used to purchase further care homes. It was also noted that in December 2007 the Southern Cross management team shared personal windfalls, totalling £36.6 million. Both the chief executive and the finance director left the company shortly after (Bowers, 2008). At the end of August 2008, Southern

Cross shareholders must have been relieved to learn that the company had negotiated a £31.1 million sale and long-term leaseback of the freeholds of nine of its homes and was in discussion about the potential sale of a further 20 freeholds (Tryhorn, 2008).

5   Residents and Staff

1. Sheldon's survey in Wolverhampton (1948) reported eight per cent who said they were very lonely and 14 per cent who said they were sometimes lonely. The figures in Townsend's survey in Bethnal Green (1957) were five per cent and 22 per cent respectively. Victor et al. (2005) reported seven per cent who said they were often or always lonely and 31 per cent who said they were sometimes lonely. All these surveys used similar single-question self-rating scales (Victor et al., 2009).

2. A national survey of care workers conducted between September 2006 and July 2007 reported that 74 per cent of those working in residential or nursing homes for all client groups (including older people) were women and that 50 per cent of the care home workforce were aged between 35 and 54 years, with a further 17 per cent aged 55 and over (Hall and Wreford, 2007, Table 127A). Ninety-two per cent were white (ibid.). Those working specifically with older people are more likely to be women and to be older (ibid., Table 12). The high proportion of women in our sample is therefore not unexpected.

3. In 2002, *The Mirror* newspaper alleged that the company that owned this home was obliging these overseas nurses to pay up to £3500 for their employment and to surrender personal documents so that they could be employed on a low wage with no chance of gaining English nurse registration (Drakeford, 2006, p. 938).

4. Information we were given on staff wages in 2005–6 showed that, as in the past, they were low. At the end of 2006, for example, a senior care worker in Pine Grange was paid £6.17 per hour, a kitchen assistant £5.89. In Laburnum House, the deputy manager's hourly wage was £9.50. As far as we know, none of these staff belonged to a union.

8   The Quality of Care

1. Since the fieldwork was completed, there have been changes to inspection procedures. Following a series of 'key' visits by CSCI inspectors in 2006–7, an overall rating of excellent, good, adequate or poor was ascribed to every care home in England. The subsequent frequency of inspection is based on this rating. With an excellent rating, inspectors expect to visit at least once every three years, good homes are inspected every two years, adequate homes every year and poor homes at least twice a year. Service providers that score adequate or poor are required to produce improvement plans detailing how and when the improvements are to be achieved. The overall ratings are based on a new set of eight outcomes, introduced in January 2008. They are quality of life, exercising choice and control, making a positive contribution, personal dignity and respect, freedom from discrimination and harassment, improved health and emotional wellbeing, economic wellbeing, leadership and management.

2. Alongside initiatives such as *No Secrets* (DH, 2000) and *In Safe Hands* (National Assembly of Wales, 2000), the Care Standards Act 2000 determined that care staff should not be allowed to start work until they had been checked as part of the CRB procedure and against the POVA list, which was implemented on a phased basis from July 2004. The POVA scheme is being replaced under the provisions of the Safeguarding Vulnerable Adults Act 2006 by a new vetting and barring scheme run by the Independent Safeguarding Authority, whose remit covers England, Wales and Northern Ireland. Registration with this new scheme will become mandatory from November 2010.
3. In applying Townsend's measure to the homes we revisited, one of the 20 homes was not included because its scores were incomplete due to lack of information on certain items.
4. Of the 20 managers of the homes visited in 2005–6, 12 reported that 50 per cent or more of the residents engaged in individual pastimes; only two reported that less than one in four residents did so. Ten homes reported that 40 per cent or more of residents engaged in group activities and a further nine reported that 20–39 per cent did so.
5. Money had been raised to buy a TV set for one of the homes Townsend visited costing £72 11s. This is equivalent at 2007 prices to £1369.92p (Officer, 2008).
6. For six homes it was not possible to obtain a score on all 37 standards because at least one standard had not been assessed for several years.
7. Although the CSCI reports are public documents, they have not been source-referenced so as to preserve the anonymity of the surviving homes. As with all the surviving homes, pseudonyms have been used in Table 8.4.

9   Revisiting and Reuse

1. Bishop (2007) undertook a secondary analysis of the data of Blaxter and Patterson (1982) and Thompson (1975). Blaxter and Patterson drew upon 58 three-generation families and focused on health issues and intergenerational transmission. Thompson examined social change in the early twentieth century through interviews with 444 people constituting a quota sample of the 1911 population. Bishop's focus was on individuated eating and the consumption of convenience foods. Savage (2007) used data from the Mass-Observation Archive comparing its correspondents' comments on social class identities in 1948 with those submitted by the M-O panel in 1990. Silva (2007) compared a 'virtual' and a 'classic' ethnography of everyday family life of 16 households with school-aged children; for the former she video-recorded daily activities, for the latter she followed the 'classic' methods of Bott (1957). She was the only one of the three to use the same sample.

10   Continuity and Change in Residential Care for Older People

1. In 2007, major providers (owners of three or more homes) were supplying over 60 per cent of places and owned over half of all homes in the private sector (Laing & Buisson, 2007).

2. To amplify this point, while we were writing this chapter, it was reported on BBC Radio 4 World at One News (7 October 2009) that the Court of Appeal had lifted an injunction prohibiting Wolverhampton Council from closing one of its care homes where a 106-year-old woman, Louisa Watts, had been living for the last five years. Despite resistance, she and nine other remaining residents were to be moved to other homes.

3. See Chapter 2, note 11.

4. See Chapter 4, note 1.

References

Allen, I., Hogg, D. and Peace, S. (1990) *Elderly People: Choice, Participation and Satisfaction* (London: Policy Studies Institute).

Age Concern and Mental Health Foundation (2006) *Inquiry into Mental Health and Well-Being in Later Life* (London: Age Concern and Mental Health Foundation).

Anchor Housing Trust (1998) *Losing a Friend to Find a Home: The Dilemma of Older People forced to Decide between Keeping their Pets and Finding a Place to Live* (Oxfordshire: Anchor Housing Trust).

Anonymous (1963) 'Declining Years', *Times Literary Supplement*, 1 March, p. 150.

Audit Commission (1986) *Making a Reality of Community Care* (London: HMSO).

—— (1998) *Home Alone: The Role of Housing in Community Care* (London: The Audit Commission).

Avebury, K. (1984) *Home Life: A Code of Practice for Residential Care* (London: Centre for Policy on Ageing).

Baines, S., Lie, M. and Wheelock, J. (2006) *Volunteering, Self-Help and Citizenship in Later Life* (Newcastle upon Tyne: Report for Age Concern Newcastle).

Baldock, C. V. (1999) 'Seniors as Volunteers: An International Perspective on Policy', *Ageing and Society*, vol. 19, part 5, pp. 581–602.

Banks, L., Haynes, P., Balloch, S. and Hill, M. (2006) *Changes in Communal Provision for Adult Social Care: 1991–2001*, JRF Findings 0416, June (York: Joseph Rowntree Foundation).

Barnes, M. and Taylor, S. (2007) *Good Practice Guide: Involving Older People in Research – Examples, Purposes and Good Practice* (ERA-AGE European Research Area in Ageing Research).

Barton, R. (1959) *Institutional Neurosis* (Bristol: John Wainwright and Sons).

Bebbington, A., Darton, R. and Netten, A. (2001) *Care Homes for Older People: Vol. 2. Admissions, Needs and Outcomes* (Canterbury: Personal Social Services Research Unit, University of Kent).

Beck, U. (1992) *Risk Society: Towards a New Modernity* (London: Sage).

Bell, C. (1977) 'Reflections on the Banbury Restudy' in C. Bell and H. Newby (eds) *Doing Sociological Research* (London: George Allen and Unwin).

Bengtson, V. L. and Schaie, K. W. S. (1999) *Handbook of Theories of Aging* (New York: Springer Publishing).

Bengtsson, A. (2004) 'Outdoor Environments for Older People in Health Facilities: A Case Study on the Experience of Accessibility', Conference Proceedings, International Association for People-Environment Studies on CD Rom, ISBN 3–85, 263–9.

Bennett, G. (1990) 'Action on Elder Abuse in the 1990s: New Definitions will Help', *Geriatric Medicine*, April, pp. 53–4.

Beresford, P. (2007a) 'The Role of Service User Research in Generating Knowledge-Based Health and Social Care: From Conflict to Contribution', *Evidence and Policy*, vol. 3, no. 3, pp. 329–41.

—— (2007b) 'User Involvement, Research and Health Inequalities: Developing New Directions', *Health and Social Care in the Community*, vol. 15, no. 4, pp. 306–12.

Beveridge, W. (1942) *Social Insurance and Allied Services* (London: HMSO).

Bhatti, M. (2006) '"When I'm in the garden I can create my own paradise": Homes and Gardens in Later Life', *The Sociological Review*, vol. 54, no. 2, pp. 318–41.

Bishop, E. (2007) 'A Reflexive Account of Reusing Qualitative Data: Beyond Primary/Secondary Dualism', *Sociological Research Online*, vol. 12, issue 3, http://www.socresonline.org.uk/12/3/2.html.

—— (2008) 'UK Data Archive: Diverse Resources for Studying Older People and Ageing', paper presented at seminar on Secondary Analysis and Re-Using Archived Data, organised by the Centre for Ageing and Biographical Studies at The Open University and the Centre for Policy on Ageing, 12 December.

Blakemore, K. and Griggs, E. (2007) *Social Policy: An Introduction* (third edition) (Maidenhead: Open University Press).

Bland, R., Bland, R., Cheetham, J., Lapsley, I. and Llewellyn, S. (1992) *Residential Homes for Elderly People: Their Costs and Quality* (Edinburgh: HMSO).

Blaxter, M. and Patterson, E. (1982) *Mothers and Daughters: A Three-generational Study of Health Attitudes and Behaviour* (London: Heinemann Educational Books).

Booth, T. (1985) *Home Truths: Old People's Homes and the Outcome of Care* (Aldershot: Gower).

Bornat, J. (2010) 'Remembering in Later Life: Generating Individual and Social Change' in D. Ritchie (ed.) *The Oxford Handbook to Oral History* (New York: Oxford University Press).

Bosanquet, N. (1978) *A Future for Old Age* (London: Temple Smith).

Bott, E. (1957) *Family and Social Network* (London: Tavistock).

Bowers, S. (2008) 'Southern Cross in Emergency Talks over £46m Loan Deadline', *The Guardian*, 1 July.

Brooker, D. and Surr, C. (2005) *Dementia Care Mapping: Principles and Practice* (Bradford: University of Bradford).

Burgner Report (1996) *The Regulation and Inspection of Social Services* (London: Department of Health).

Butler, A., Oldman, C. and Greve, J. (1983) *Sheltered Housing for the Elderly: Policy, Practice and the Consumer* (London: George Allen and Unwin).

Bytheway, B. (2005) 'Age-Identities and the Celebration of Birthdays', *Ageing & Society*, vol. 25, part 4, pp. 463–78.

Bytheway, B., Ward, R., Holland, C. and Peace, S. (2007) *Too Old: Older People's Accounts of Discrimination, Exclusion and Rejection*, A report from the Research on Age Discrimination Project (RoAD) to Help the Aged (London: Help the Aged).

Canal, G. O. (2004) 'Photography in the Field: Word and Image in Ethnographic Research' in S. Pink, L. Kurti and A. I. Afonso (eds) *Working Images: Visual Research and Representation in Ethnography* (London: Routledge).

Carstairs, V. and Morrison, N. (1971) *The Elderly in Residential Care* (Edinburgh: Scottish Home and Health Department).

Carvel, J. (2007) 'Prospect of Moving into a Care Home Frightens Two Thirds of Britons', *Society Guardian*, 3 December.

Chalfont, G. (2005) 'Creating Enabling Outdoor Environments for Residents', *Nursing and Residential Care*, vol. 7, no. 10, pp. 454–7.

—— (2007) *Design for Nature in Dementia Care* (London: Jessica Kingsley).

—— (2008) 'The Dementia Care Garden: Innovation in Design and Practice', *Journal of Dementia Care*, January–February, pp. 18–20.

Chalfont, G. and Rodiek, S. (2005) 'Building Edge: An Ecological Approach to Research and Design of Environments for People with Dementia', *Alzheimer's Care Quarterly*, vol. 6, no. 4, pp. 341–8.

Chaplin, E. (2006) 'Photographs in Social Research: The Residents of South London Road' in P. Hamilton (ed.) *Visual Research Methods, Volume 1V* (London: Sage).

Charles, N., Davies, C. A. and Harris, C. (2008) *Families in Transition: Social Change, Family Formation and Kin Relationships* (Bristol: The Policy Press).

Clare, L., Bruce, E., Surr, C. and Downs, M. (2008) 'The Experience of Living with Dementia in Residential Care: An Interpretative Phenomenological Analysis', *The Gerontologist*, vol. 48, pp. 711–20.

Clough, R. (1981) *Old Age Homes* (London: George Allen and Unwin).

—— (1982) *Residential Work* (London: Macmillan).

—— (ed.) (1996) *The Abuse of Care in Residential Institutions* (London: Whiting and Birch).

—— (2008) 'Home Improvement', *The Guardian (Society)*, 9 January.

Clough, R., Green, G., Hawkes, B., Raymond, G. and Bright, L. (2006) *Older People as Researchers: Evaluating a Participative Project* (York: Joseph Rowntree Foundation).

Clough, R., Leamy, M., Miller, V. and Bright, L. (2004) *Housing Decisions in Later Life* (Basingstoke: Palgrave Macmillan).

Coates, D. (2005) *Prolonged Labour: The Slow Birth of New Labour Britain* (Basingstoke: Palgrave Macmillan).

Cohen, S. and Taylor, L. (1972) *Psychological Survival: The Effects of Long-Term Imprisonment* (London: Allen Lane).

Community Care (2007) 'Care Homes', *Community Care*, issue 1693, 4 October.

Corbett, J. (1997) 'Provision of Prescribing Advice for Nursing and Residential Home Patients', *Pharmaceutical Journal*, vol. 259, pp. 422–4.

Corti, L. and Thompson, P. (2004) 'Secondary Analysis of Archived Data' in C. Seale, G. Giampieto, J. F. Gubrium and D. Silverman (eds) *Qualitative Research Practice* (London: Sage).

Crossman, R. H. S. (1962) 'Old People', *New Statesman*, vol. LXIV, no. 1659, pp. 930–1.

Croucher, K. and Rhodes, P. (2006) *Paying for Long Term Care: Moving Forward* (York: Joseph Rowntree Foundation).

Croucher, K., Hicks, L. and Jackson, K. (2006) *Housing with Care for Later Life: A Literature Review* (York: Joseph Rowntree Foundation).

Crow, G., Wiles, R., Heath, S. and Charles, V. (2006) 'Research Ethics and Data Quality: The Implications of Informed Consent', *International Journal of Research Methodology*, vol. 9, no. 2. pp. 83–95.

CSCI (Commission for Social Care Inspection) (2007) *Safe as Houses: What Drives Investment in Social Care?* (London: CSCI).

—— (2008) *The State of Social Care in England 2006–07* (London: CSCI).

—— (2009) *The State of Social Care in England 2007–08* (London: CSCI).

Cwmni Iaith on behalf of the Welsh Language Board (2002) *An Overview of the Welsh Language Provision in Care Homes for Older People in Wales' Eight Most Welsh Speaking Counties*, http://www.bwrdd-yr-iaith.org.uk/download.php/pID=3206.4. Accessed 15 January 2007.

Dalley, G., Unsworth, L., Keightley, D., Waller, M., Davies, T. and Morton, R. (2004) *How Do We Care? The Availability of Registered Care Homes and Children's Homes in England and their Performance against National Minimum Standards 2002–03* (London: The Stationery Office).

Darton, R. (2004) 'What Types of Home are Closing? The Characteristics of Homes which Closed between 1996 and 2001', *Health and Social Care in the Community*, vol. 12, no. 3, pp. 254–64.

Darton, R., Bäumber, T., Callaghan, L., Holder, J., Netten, A. and Towers, A.-M. (2008) *Evaluation of Extra Care Housing Funding Initiative: Initial Report*, PSSRU Discussion Paper 2506/2 (Canterbury: PSSRU, University of Kent).

Darton, R. and Wright, K. (1990) 'The Characteristics of Non-Statutory Residential and Nursing Homes' in R. Parry (ed.) *Privatisation*, Research Highlights in Social Work 18 (London: Jessica Kingsley).

Davies, B. P. and Knapp, M. R. J. (1981) *Old People's Homes and the Production of Welfare* (London: Routledge and Kegan Paul).

Davies, C. A. (1999) *Reflexive Ethnography* (London: Routledge).

Davies, C. A. and Charles, N. (2002) 'The Piano in the Parlour: Methodological Issues in the Context of a Restudy', *Sociological Research Online*, vol. 7, no. 2. Available at http://www.socresonline.org.uk/7/2/davies.html.

Day, P., Klein, R. and Redmayne, S. (1996) *Why Regulate? Regulating Residential Care for Elderly People* (Bristol: The Policy Press).

DH (Department of Health) (1997) *The New NHS: Modern, Dependable* (London: The Stationery Office).

—— (1998) *Modernising Social Services: Promoting Independence, Improving Protection, Raising Standards* (London: The Stationery Office).

—— (1999) *Fit for the Future? National Required Standards for Residential and Nursing Homes for Older People* (London: Department of Health).

—— (2000) *No Secrets: Guidance on Developing and Implementing Multi-Agency Policies and Procedures to Protect Vulnerable Adults from Abuse* (London: Department of Health).

—— (2001a) *Care Homes for Older People: National Minimum Standards* (London: The Stationery Office).

—— (2001b) *National Service Framework for Older People* (London: Department of Health).

—— (2002) *National Minimum Standards for Care Homes for Older People/National Minimum Standards for Care Homes for Younger Adults (18–65): Proposed Amended Environmental Standards* (London: Department of Health).

—— (2004) *Choosing Health: Making Healthy Choices Easier* (London: The Stationery Office).

—— (2005a) *Independence, Well-Being and Choice: Our Vision for the Future of Social Care for Adults in England* (London: The Stationery Office).

—— (2005b) *Research Governance Framework for Health and Social Care* (second edition) (London: Department of Health).

—— (2006) *Our Health, Our Care, Our Say: A New Direction for Community Services* (London: The Stationery Office).

—— (2008) *Transforming Social Care*, Local Authority Circular 1, 17 January.

—— (2009) *Shaping the Future of Care Together* (London: The Stationery Office).

DHSS (Department of Health and Social Security) (1981) *Growing Older* (London: HMSO).

DHSS/Welsh Office (1973) *Residential Accommodation for Elderly People*, Local Authority Building Note No. 2 (London: HMSO).

—— (1978) *A Happier Old Age* (London: HMSO).

DH/Welsh Office (1995) *Moving Forward: A Consultation Document on the Regulation and Inspection of Social Services* (London: Department of Health).

Doll, R. and Hill, A. B. (1954) 'The Mortality of Doctors in Relation to their Smoking Habits: A Preliminary Report', *British Medical Journal*, vol. 228, pp. 1451–5.

Downs, M. (1997) 'The Emergence of the Person in Dementia Research', *Ageing and Society*, vol. 17, part 5, pp. 597–608.

Drake, M. (2005) 'Inside-Out or Outside-In? The Case of Family and Local History' in R. Finnegan (ed.) *Participating in the Knowledge Society: Researchers Beyond the University Walls* (Basingstoke: Palgrave Macmillan).

Drakeford, M. (2006) 'Ownership, Regulation and the Public Interest: The case of Residential Care for Older People', *Critical Social Policy*, issue 26, no. 4, pp. 932–44.

Duffin, C. (2008) 'Designing Care Homes for People with Dementia', *Nursing Older People*, vol. 20, no. 4, pp. 22–4.

Dutton, R. (2009) *'Extra Care' Housing and People with Dementia: A Scoping Review of the Literature 1998–2008* (Housing 21 on behalf of the Housing and Dementia Research Consortium).

Elkan, R. and Kelly, D. (1991) *A Window in Homes: Links between Residential Care Homes and the Community – A Literature Review* (Surbiton: Social Care Association).

ESDS (Economic and Social Data Service) (nda) 'Identifiers and anonymisation: Dealing with confidentiality', ESDS Access and Preservation, http://www.esds. ac.uk/aandp/create/identguideline.asp. Accessed 3 March 2006.

—— (ndb) 'End user licence', ESDS Access and Preservation, http://www.esds. ac.uk/aandp/access/licence.asp. Accessed 3 March 2006.

Estes, C. (1979) *The Aging Enterprise* (San Francisco: Jossey-Bass).

Estes, C., Biggs, S. and Phillipson, C. (2003) *Social Theory, Social Policy and Ageing: A Critical Introduction* (Maidenhead: Open University Press).

Evans, S. (2009) Personal Communication, 18 October.

Evans, G., Hughes, B. and Wilkin, D. with Jolley, D. (1981) *The Management of Mental and Physical Impairment in Non-specialist Homes for the Elderly*, Research Report No. 4 (Manchester: Departments of Psychiatry and Community Medicine, University of Manchester).

Evans, S. and Vallelly, S. (2007) *Social Well-Being in Extra Care Housing* (York: Joseph Rowntree Foundation).

Evans, T. and Thane, P. (2006) 'Secondary Analysis of Dennis Marsden *Mothers Alone*', *Methodological Innovation Online* [Online], vol. 1, no. 2, http://erdt. plymouth.ac.uk/mionline/public_html/viewarticle.php?id=31.

Fielding, N. G. and Fielding, J. L. (2008) 'Resistance and Adaptation to Criminal Identity: Using Secondary Analysis to Evaluate Classic Studies of Crime and Deviance', *Historical Social Research*, vol. 33, no. 3, pp. 75–93.

Finnegan, R. (1992) *Oral Traditions and the Verbal Arts: A Guide to Research Practice* (London: Routledge).

Finnegan, R. (ed.) (2005) *Participating in the Knowledge Society: Researchers beyond the University Walls* (Basingstoke: Palgrave Macmillan).

Firth Committee (1987) *Public Support for Residential Care: Report of a Joint Central and Local Government Working Party* (London: DHSS).

Froggatt, K. and Payne, S. (2006) 'A Survey of End of Life Care in Care Homes: Issues of Definition and Practice', *Health and Social Care in the Community*, vol. 14, no. 4, pp. 341–8.

Froggatt, K., Davies, S. and Meyer, J. (eds) (2008) *Understanding Care Homes: A Research and Development Perspective* (London: Jessica Kingsley).

Ganzel, B. (1984) *Dust Bowl Descent* (Lincoln/London: University of Nebraska Press).

Garner, J. and Evans, S. (2000) *Institutional Abuse of Older Adults* (London: Royal College of Psychiatrists).

Geertz, C. (1988) *Works and Lives: The Anthropologist as Author* (Cambridge: Polity Press).

Gentleman, A. (2009) 'Dying Days', *The Guardian G2*, 14 July, pp. 1, 5–9.

Gibson, F. (1994) *Reminiscence and Recall* (London: Age Concern England).

Giddens, A. (1994) 'Living in a Post-Traditional Society' in U. Beck, A. Giddens and S. Lash (eds), *Reflexive Modernization: Politics and Tradition in the Modern Social Order* (Cambridge: Polity Press).

Gilleard, C. and Higgs, P. (2000) *Cultures of Ageing: Self, Citizen and the Body* (Harlow: Prentice-Hall).

—— (2005) *Contexts of Ageing: Class, Cohort and Community* (Cambridge: Polity Press).

—— (2007) 'Ageing without Agency: Theorizing the Fourth Age', paper presented at the 60th Annual Scientific Conference of the Gerontological Society of America, 15–17 November (San Francisco, CA).

Glendinning, C., Challis, D., Fernandez, J. L., Jacobs, S., Jones, K., Knapp, M., Manthorpe, J., Moran, N., Netten, A., Stevens, M. and Wilberforce, M. (2008) *Evaluation of the Individual Budgets Pilot Programme: Summary Report* (York: Social Policy Research Unit, University of York).

Gluck, S. B. and Patai, D. (eds) (1991) *Women's Words: The Feminist Practice of Oral History* (London: Routledge).

Goffman, E. (1961) *Asylums: Essays on the Social Situation of Mental Patients and Other Inmates* (New York: Anchor Books, Doubleday & Co.).

Goodman, C. and Woolley, R. (2004) 'Older People in Care Homes and the Primary Care Nursing Contribution: A Review of Relevant Research', *Primary Health Care Research and Development*, vol. 5, issue 3, pp. 211–18.

Goodwin, J. (2005) *From Young Workers to Older Workers: Reflections on Work in the Life Process*, End of Award Report to the ESRC, http://www.esrcsocietytoday.ac.uk/

Guillemard, A. M. (1980) *La Viellesse et L'Etat [Old Age and the State]* (Paris: Press Universitaires de France).

Gupta, H. (1980) 'Group Living in Residential Homes for Elderly People', *Choosing How to Live* (London: MIND).

Hall, D. and Bytheway, B. (1982) 'The Blocked Bed: Definition of the Problem', *Social Science and Medicine*, vol. 16, no. 22, pp. 1985–91.

Hall, L. and Wreford, S. (2007) *National Survey of Care Workers: Final Report* (Leeds: Skills for Care).

Hall, S. (1973) 'The Determination of News Photographs' in S. Cohen and J. Young (eds) *The Manufacture of News, Social Problems, Deviance and the Mass Media* (London: Constable).

Hancock, R., Askham, J., Nelson, H. and Tinker, A. (1999) *Home Ownership in Later Life: Financial Benefit or Burden?* (York: Joseph Rowntree Foundation).

Hanson, J. (1972) *Residential Care Observed* (London: Age Concern and National Institute for Social Work Training).

Harper, D. (2001) *Changing Works: Visions of a Lost Agriculture* (Chicago: University of Chicago Press).

Harris, A. (1968) *Social Welfare for the Elderly: Volume1 – Comparison of Areas and Summary*, Government Social Survey (London: HMSO).

Health Advisory Service (1982) *The Rising Tide: Developing Services for Mental Illness in Old Age* (London: HMSO).

Heaton, J. (2004) *Reworking Qualitative Data* (London: Sage).

Help the Aged (2008) *Residential Care Briefing 2008* (London: Help the Aged).

Hendrickson, P. (2007) *Bound for Glory: America in Colour 1939–43* (Harry Abrams in Association with the Library of Congress: New York).

Henry, J. (1963) *Culture Against Man* (New York: Random House).

Hickley, M. and Greenhill, S. (2008) 'Ancient Britain: For the First Time in History, there are More OAPs than Children', *Mail Online*, 22 August. Accessed 22 February 2009, http://www.dailymail.co.uk/home/search.html?searchPhrase= Ancient+Britain.

Higgins, J. (1989) 'Defining Community Care: Realities and Myths', *Social Policy and Administration*, vol. 23. no. 1, pp. 3–16.

Hitch, D. and Simpson, A. (1972) 'An Attempt to Assess a New Design in Residential Homes for the Elderly', *British Journal of Social Work*, vol. 2, no. 1, pp. 481–50.

HM Government (2007a) *Putting People First: A Shared Vision and Commitment to the Transformation of Adult Social Care* (London: HM Government).

—— (2007b) *Building on Progress: Public Services* (London: Prime Minister's Strategy Unit, Cabinet Office).

—— (2008) *The Case for Change – Why England Needs A New Care and Support System* (London: Department of Health).

Holden, C. (2002) 'British Government Policy and the Concentration of Ownership in Long-Term Care Provision', *Ageing & Society*, vol. 22, part 1, pp. 79–94.

House of Commons Health Committee (2002) *Delayed Discharges*, House of Commons Paper 617-I, Third Report (Session 2001–2) (London: The Stationery Office).

Howe, G. (1969) *Report of the Committee of Inquiry into Allegations of Ill-treatment of Patients and Other Irregularities at Ely Hospital, Cardiff* (London: HMSO).

Hubbard, G., Downs, M. and Tester, S. (2002) 'Including the Perspective of Older People in Institutional Care Including Residents unable to give Informed Consent during the Consent Process' in H. Wilkinson (ed.) *The Perspectives of People with Dementia: Research Methods and Motivations* (London: Jessica Kingsley).

Inskip, J. H. (1974) *Report of the Committee of Inquiry into South Ockendon Hospital* (London: HMSO).

Jack, R. (ed.) (1998) *Residential versus Community Care: The Role of Institutions in Welfare Provision* (Basingstoke: Palgrave Macmillan).

Jenks, M. (1978) 'A Case Study of Two Old People's Homes', *Architect's Journal*, vol. 24, May.

Johnson, J. (1982) 'Two Residential Homes for the Elderly: A Comparative Study', MA thesis (University of Keele: Unpublished).

—— (1993) 'Does Group Living Work?' in J. Johnson and R. Slater (eds) *Ageing and Later Life* (London: Sage).

—— (2002) 'Taking Care of Later Life: A Matter of Justice?' *British Journal of Social Work*, vol. 32, no. 6, pp. 739–50.

Johnson, J., Rolph, S. and Smith, R. (2006) 'Drawing on Expertise and Experience: Older People as Co-Researchers', paper presented at the annual conference of the British Society of Gerontology, University of Bangor, 7–9 September.

—— (2007) 'Revisiting "The Last Refuge": Present-Day Methodological Challenges' in M. Bernard and T. Scharf (eds) *Critical Perspectives on Ageing Societies* (Bristol: The Policy Press).

—— (2008) '"The Last Refuge" Revisited: A Case Study', *Generations Review*, vol. 18, no. 1. Available at http://www.britishgerontology.org/08newsletter1/research4.asp.

—— (2010) 'Uncovering History: Private Sector Homes for Older People', *Journal of Social Policy*, vol. 39, part 2, pp. 235–54.

Johnson, M. (1983) 'Controlling the Cottage Industry', *Community Care*, 25 August, pp. 16–18.

Jones, K. (1962) 'The Workhouse is Still with Us', *New Society*, no. 10, 6 December, pp. 30–1.

—— (1967) 'The Development of Institutional Care' in *New Thinking about Institutional Care* (London: British Association of Social Workers).

Jordanova, L. (2000) *History in Practice* (London: Arnold).

Judd, S., Marshall, M. and Phippen, P. (1998) *Design for Dementia* (London: Hawker Publications).

Judge, K. and Sinclair, I. (1986) *Residential Care for Elderly People: Research Contributions to the Development of Policy and Practice* (London: HMSO).

Karn, V. (1977) *Retiring to the Seaside* (London: Routledge and Kegan Paul).

Katz, S. (1996) *Disciplining Old Age: The Formation of Gerontological Knowledge* (Charlottesville: University Press of Virginia).

Katz, J. S. and Peace, S. (eds) (2003) *End of Life in Care Homes* (Oxford: Oxford University Press).

Kay, D. W., Beamish, P. and Roth, M. (1964) 'Some Medical and Social Characteristics of Elderly People under State Care: A Comparison of Geriatric Wards, Mental Hospitals and Welfare Homes', *British Journal of Psychiatry*, vol. 110, pp. 146–58.

Kellaher, L. (2000) *A Choice Well Made: Mutuality as a Governing Principle in Residential Care* (London: Centre for Policy on Ageing).

Kellaher, L., Peace, S., Weaver, T. and Willcocks, D. (1988) *Coming to Terms with the Private Sector: Regulatory Practice for Residential Care Homes and Elderly People* (London: CESSA, The Polytechnic of North London).

Kemp, J. (2005) 'Links in the Chain', *'The Last Refuge' Revisited. Newsletter No. 1*, October, http://www.open.ac.uk/hsc/lastrefuge/newsletters.htm.

Killick, J. and Allan, K. (2001) *Communication and the Care of People with Dementia* (Buckingham: Open University Press).

Kitwood, T. (1997) *Dementia Reconsidered: The Person Comes First* (Buckingham: Open University Press).

Kitwood, T. and Bredin, K. (1992) 'Towards a Theory of Dementia Care: Personhood and Wellbeing', *Ageing and Society*, vol. 12, part 3, pp. 269–87.

Knapp, M., Hardy, B. and Forder, J. (2001) 'Commissioning for Quality: Ten Years of Social Care Markets in England', *Journal of Social Policy*, vol. 30, part 2, pp. 283–306.

Kynaston, D. (2007) *Austerity in Britain, 1945–1951* (London: Bloomsbury).

Laing & Buisson (2007) *Care of Elderly People UK Market Survey 2007* (twentieth edition) (London: Laing & Buisson).

—— (2008) 'Care of Elderly and Physically Disabled People', *Healthcare Market Review 2007–2008* (London: Laing & Buisson).

—— (2009) *Laing's Healthcare Market Review 2008–9* (twenty-first edition) (London: Laing & Buisson).

Law, C. M. and Warnes, A. M. (1973) 'The Movement of Retired People to Seaside Resorts: A Study of Morecambe and Llandudno', *Town Planning Review*, vol. 44, pp. 373–90.

Lee-Treweek, G. (1994) 'Bedroom Abuse: The Hidden Work in a Nursing Home', *Generations Review*, vol. 4, no. 1, pp. 2–4.

Lewis, C. A. (1973), 'People–Plant Interaction: A new Horticultural Perspective', *American Horticulturalist*, vol. 52, pp. 18–25.

Lewis, L. A. (2004) 'Modesty and Modernity: Photography, Race and Representation on Mexico's Costa Chica (Guerrero)', *Identities: Global Studies in Culture and Power*, vol. 11, pp. 471–99.

Leybourne-White, G. and White, K. (1945) *Children for Britain* (London: Pilot Press).

Lincoln, Y. S. and Guba, E. G. (1985) *Naturalistic Inquiry* (Newbury Park: Sage).

Lipman, A. and Slater, R. (1976) 'Homes for Old People: Towards a Positive Environment', *The Gerontologist*, vol. 25, no. 2, pp. 147–56.

—— (1977) 'Status and Social Appropriation in Eight Homes for Old People', *The Gerontologist*, vol. 17, no. 3, pp. 250–5.

Lipman, A., Slater, R. and Harris, H. (1979) 'The Quality of Verbal Interaction in Homes for Old People', *Gerontology*, vol. 25, no. 5, pp. 275–84.

Lombard, D. (2008) 'Bolton Pressures for Residential Spending Reduction', *Community Care*, 1 May, p. 11.

Lowes, L. and Hulatt, I. (eds) (2005) *Involving Service Users in Health and Social Care Research* (London: Routledge).

Lundgren, E. (2000) 'Homelike Housing for Elderly People – Materialized Ideology', *Housing, Theory and Society*, vol. 17, pp. 109–20.

Mackintosh, S., Means, R. and Leather, P. (1990) *Housing in Later Life* (Bristol: School of Advanced Urban Studies, University of Bristol).

Macleod, J. and Thomson, R. (2009) *Researching Social Change* (London: Sage).

Magnus, G. (2009) 'Dependency Time-Bomb', *The Guardian*, 4 February, p. 28.

Manthorp, C. (2008) 'Up the Garden Path to Contented Care', *Society Guardian*, 13 August.

Manthorpe, J. (2008) 'Dementia Care Quality in Homes', *Community Care*, issue 1720, 1 May.

Marsden, D. (1969) *Mothers Alone: Poverty and the Fatherless Family* (London: Allen Lane, The Penguin Press).

Marshall, M. (2001) 'Dementia and Technology', in S. M. Peace and C. A. Holland (eds) *Inclusive Housing in an Ageing Society* (Bristol: The Policy Press).

—— (ed.) (2004) *Perspectives on Rehabilitation and Dementia* (London: Jessica Kingsley).

Mathiason, N. (2007) 'Why Granny is a Profit Centre', *The Observer*, 4 November.

McNabb, M. (2008) 'First Things not First', *The Social Care Experts Blog*, http://www.communitycare.co.uk/blogs/social-care-experts-blog/2008/08/first-things-not-first.html.

Meacher, M. (1972) *Taken for a Ride: Special Residential Homes for Confused Old People: A Study of Separatism in Social Policy* (London: Longman).

Means, R. and Smith, R. (1983) 'From Public Assistance Institutions to "Sunshine Hotels": Changing State Perceptions about Residential Care for Elderly People', *Ageing and Society*, vol. 3, part 2, pp. 157–81.

—— (1985) *The Development of Welfare Services for Elderly People* (London: Croom Helm).

—— (1994) *Community Care: Policy and Practice* (London: Macmillan).

—— (1998) *From Poor Law to Community Care: The Development of Welfare Services for Elderly People 1939–1971* (Bristol: The Policy Press).

Means, R., Morbey, H. and Smith, R. (2002) *From Community Care to Market Care?* (Bristol: The Policy Press).

Means, R., Richards, S. and Smith, R. (2008) *Community Care: Policy and Practice* (fourth edition) (Basingstoke: Palgrave Macmillan).

Miller, E., Cook, A., Alexander, H., Cooper, S.-A., Hubbard, G., Morrison, J. and Petch, A. (2006) 'Challenges and Strategies in Collaborative Working with Service User Researchers: Reflections from the Academic Researcher', *Research, Policy and Planning*, vol. 24, no. 3, pp. 197–208.

Miller, E. J. and Gwynne, G. (1972) *A Life Apart: Pilot Study of Residential Institutions for the Physically Handicapped and the Young Chronic Sick* (London: Tavistock Publications).

Ministry of Health (1947) *National Assistance Bill – Future of Public Assistance Institutions*, Circular 172/47, 11 December.

—— (1962) *Development of Local Authority Health and Welfare Services: Ten Year Plans*, Circular 2/62.

—— (1966) *Annual Report of the Ministry of Health for the year 1965* (London: HMSO).

Ministry of Housing and Local Government (1962a) *Some Aspects of Designing for Old People* (London: HMSO).

—— (1962b) *Old People's Flatlets in Stevenage* (London: HMSO).

—— (1966) *Grouped Flatlets for Old People* (London: HMSO).

Minkler, M. and Cole, T. (1991) 'Political and Moral Economy: Not Such Strange Bedfellows' in M. Minkler and C. Estes (eds) *Critical Perspectives on Aging: The Political and Moral Economy of Growing Old* (Amityville: Baywood Publishing).

Minkler, M. and Estes, C. (eds) (1991) *Critical Perspectives on Aging: The Political and Moral Economy of Growing Old* (Amityville: Baywood Publishing).

Moore, N. (2007) '(Re)Using qualitative data', *Sociological Research Online*, vol. 12, issue 3.

Morgan, D., Reed, J. and Palmer, A. (1997) 'Moving from Hospital into a Care Home: The Nurse's Role in Supporting Older People', *Journal of Clinical Nursing*, vol. 6, no. 6, pp. 463–71.

Morris, P. (1969) *Put Away: A Sociological Study of Institutions for the Mentally Retarded* (London: Routledge and Kegan Paul).

Mullan, P. (2000) *The Imaginary Time Bomb: Why an Ageing Population is Not a Social Problem* (London: I. B. Tauris).

Myles, J. (1984) *Old Age and the Welfare State* (Lawrence: University of Kansas Press).

National Assembly of Wales (2000) *In Safe Hands: Implementing Adult Protection Procedures in Wales* (Cardiff: National Assembly of Wales).

Neill, J., Sinclair, I., Gorbach, P. and Williams, J. (1988) *A Need for Care? Elderly Applicants for Local Authority Homes*, NISW Research Report (Aldershot: Avebury).

Nell, D., Alexander, A., Shaw, G. and Bailey, A.R. (2009) 'Investigating Shopper Narratives of the Supermarket in Early Post-War England, 1945–75', *Oral History*, vol. 37, no. 1, pp. 61–73.

Netten, A. (1993) *A Positive Environment?* (Aldershot: Ashgate).

Netten, A., Bebbington, A., Darton, R. and Forder, J. (2001) *Care Homes for Older People: Vol 1. Facilities, Residents and Costs* (Canterbury: Personal Social Services Research Unit, University of Kent).

Netten, A., Williams, J. and Darton, R. (2005) 'Care-Home Closures in England: Causes and Implications', *Journal of Social Policy*, vol. 25, part 3, pp. 319–38.

Nolan, M., Hanson, E., Grant, G. and Keady, J. (eds) (2007) *User Participation in Health and Social Care Research: Voices, Values and Evaluation* (Maidenhead: Open University Press).

Nolan, M. R., Davies, S., and Grant, G. (eds) (2001) *Working with Older People and Their Families: Key Issues in Policy and Practice* (Buckingham: Open University Press).

Number10.gov.uk (2008) Transcript of a speech given by the Prime Minister at the King's Fund debate on the Future of Social Care, 12 May, www.number 10-gov.uk/output/Page15496.asp.

O'Connor, H. and Goodwin, J. (2008) 'Data from the Attic – Revisiting a Lost Study: Exploring issues of Revisiting a Study Carried Out in the 1960s', paper presented to the National Conference of the Centre for Research on Families and Relationships, *Understanding Families and Relationships Over Time*, Edinburgh, 30 October.

Office of Fair Trading (2005) *Care Homes for Older People in the UK: A Market Study*, OFT Report 780 (London: Office of Fair Trading).

Office of Public Sector Information (2007) *Public Health England: The Smoke-free (Exemptions and Vehicles) Regulations 2007*, Statutory Instrument No. 765, Part 2, paragraph 5(2)(a).

Officer, Lawrence H. (2008) 'Purchasing Power of British Pounds from 1264 to 2007', *Measuring Worth*, 2008. http://www.measuringworth.com/ppoweruk/. Accessed 5 May 2009.

ONS (Office for National Statistics) (2005) All People in Communal Establishments: Census 2001, National Report for England and Wales – Part 2 (London: ONS).

—— (2006) Interim Life Tables 2003–05, Table 2 (London: ONS).

—— (2007) *Social Trends* No. 37 (Basingstoke: Palgrave Macmillan).

O'Rand, A. M. (2006) 'Stratification and the Life Course. Life-Course Capital, Life-Course Risks, and Social Inequality' in R. L. Binstock and L. K. George (eds) *Handbook of Aging and the Social Sciences* (sixth edition) (San Diego, CA: Academic Press).

Page, R. (2002) 'Peter Townsend – Still Going Strong', *SPA News*, May/June, pp. 1–5.

Parker, C., Barnes, S., McKee, K., Morgan, K., Torrington, J. and Tregenza, P. (2004) 'Quality of Life and Building Design in Residential and Nursing Homes for Older People', *Ageing & Society*, vol. 24, part 6, pp. 941–62.

Parry, J. and Taylor, R. F. (2007) 'Orientation, Opportunity and Autonomy: Why People Work after State Pension Age in Three Areas of England', *Ageing & Society*, vol. 27, part 4, pp. 579–98.

Passini, R. (1992) *Wayfinding in Architecture* (Chichester: John Wiley).

Payne, R. (1972) *Report of the Committee of Inquiry into Whittingham Hospital* (London: HMSO).

Peace, S. (ed.) (1999) *Involving Older People in Research* (London: Centre for Policy on Ageing).

Peace, S. (2002) 'The Role of Older People in Research' in A. Jamieson and C. R. Victor (eds), *Researching Ageing and Later Life* (Buckingham: Open University Press).

Peace, S. and Holland, C. (2001) 'Homely Residential Care: A Contradiction in Terms?' *Journal of Social Policy*, vol. 30, part 3, pp. 393–410.

Peace, S. and Johnson, J. (1998) 'Living Arrangements of Older People' in M. Bernard and J. Phillips (eds), *Social Policy and Older People* (London: Centre for Policy on Ageing).

Peace, S. and Kellaher, L. (1993) 'Rest Assured' in J. Johnson and R. Slater (eds) *Ageing and Later Life* (London: Sage).

Peace, S., Holland, C. and Kellaher, L. (2006) *Environment and Identity in Later Life* (Maidenhead: Open University Press).

Peace, S., Kellaher, L. and Willcocks, D. (1982) *A Balanced Life?* (London: Survey Research Unit, Polytechnic of North London).

—— (1997) *Re-Evaluating Residential Care* (Buckingham: Open University Press).

Pendleton, A. (2008) 'Putting Acton Square on the Map', *'The Last Refuge' Revisited. Newsletter No. 1*, March, http://www.open.ac.uk/hsc/lastrefuge/newsletters.htm.

Percival, J. (2002) 'Domestic Spaces: Uses and Meanings in the Daily Lives of Older People', *Ageing & Society*, vol. 22, part 6, pp. 729–49.

Phillips, D. R, Vincent, J. V. and Blacksell, S. (1988) *Home from Home? Private Residential Care for Elderly People* (Sheffield: Joint Unit for Social Services Research, University of Sheffield/Community Care).

Phillipson, C. (1982) *Capitalism and the Construction of Old Age* (London: Macmillan).

Phillipson, C., Bernard, M., Phillips, J. and Ogg, J. (1998) 'The Family and Community Life of Older People: Household Composition and Social Networks in Three Urban Areas', *Ageing and Society*, vol. 18, part 3, pp. 259–90.

—— (2001) *The Family and Community Life of Older People: Social Networks and Social Support in Three Urban Areas* (London: Routledge).

Philpot, T. (2003a) *On The Homes Front: The Catholic Church and Residential Care for Older People* (London: Caritas-Social Action).

—— (2003b) 'Catholic Trends', *Community Care*, 13–19 March, pp. 36–7.

Pickard, S. (2009) 'Governing Old Age: The "Case Managed" Older Person', *Sociology*, vol. 43, no. 1, pp. 67–84.

Pincus, A. (1968) 'The Definition and Measurement of the Institutional Environment in Homes for the Aged', *The Gerontologist*, vol. 8, no. 3, pp. 207–10.

Prosser, J. and Schwartz, D. (1998) 'Photographs within the Sociological Research Process' in J. Prosser (ed.) *Image-Based Research: A Sourcebook for Qualitative Researchers* (London: RoutledgeFalmer).

Quadagno, J. (1988) *Transformation of Old Age Security* (Chicago: University of Chicago Press).

Ray, M. (2007) 'Redressing the Balance? The Participation of Older People in Research' in M. Bernard and T. Scharf (eds) *Critical Perspectives on Ageing Societies* (Bristol: The Policy Press).

Reed, J. and Roskell-Payton, V. (1995) 'Accomplishing Friendships in Nursing and Residential Homes', paper presented at the Third European Congress of Gerontology, September (Amsterdam).

Reed, J., Roskell-Payton, V. and Bond, S. (1998) 'The Importance of Place for Older People Moving into Care Homes', *Social Science & Medicine*, vol. 46, issue 7, pp. 859–67.

Rees Jones, I., Hyde, M., Victor, C., Wiggins, R. D., Gilleard, C. and Higgs, P. (2008) *Ageing in a Consumer Society: From Passive to Active Consumption* (Bristol: The Policy Press).

Rieger, J. H. (1996) 'Photographing Social Change', *Visual Sociology*, vol. 11, no. 1, pp. 5–49.

Riley, M. W., Foner, A. and Riley, J. W. (1999) 'The Aging and Society Paradigm' in V. L. Bengtson and K. Warner Schaie (eds) *Handbook of Theories of Aging* (New York: Springer Publishing).

Riseborough, M. and Fletcher, P. (2008) *Extra Care Housing: What is it?* (revised edition) (London: Housing Learning and Improvement Network).

Robb, B. (1967) *Sans Everything: A Case to Answer* (London: Nelson).

Robertson, A. (1990) 'The Politics of Alzheimer's Disease: A Case Study in Apocalyptic Demography', *International Journal of Health Services*, vol. 20, no. 3, pp. 429–42.

Robertson, A. (1997) 'Beyond Apocalyptic Demography: Towards a Moral Economy of Interdependence', *Ageing and Society*, vol. 17, part 4, pp. 425–46.

Rolph, S. (2000) 'The History of Community Care for People with Learning Difficulties in Norfolk, 1930–1980: The Role of Two Hostels', PhD thesis (The Open University: Unpublished).

Rolph, S., Johnson, J. and Smith, R. (2009) 'Using Photography to Understand Change and Continuity in the History of Residential Care for Older People', *International Journal of Social Research Methodology*, vol. 12, no. 5, pp. 421–39, http://www.informaworld.com.

Rose, G. (2007) *Visual Methodologies: An Introduction to the Interpretation of Visual Materials* (London: Sage).

Rosser, C. and Harris, C. (1965) *The Family and Social Change* (London: Routledge and Kegan Paul).

Rowntree, B. S. (1947) *Old People: A Report of a Survey Committee on the Problems of Ageing and the Care of Old People* (London: Oxford University Press).

Royal Commission on Population (1949) *The Report of the Royal Commission on Population* (London: HMSO).

Royal Commission on the Law Relating to Mental Illness and Mental Deficiency (1957) *Report* (London: HMSO).

Savage, M. (2007) 'Changing Social Class Identities in Post-War Britain: Perspectives from Mass-Observation', *Sociological Research Online*, vol. 12, issue 3, http://www.socresonline.org.uk/12/3/6.html.

Schweitzer, P. (1998) *Reminiscence in Dementia Care* (London: Age Exchange).

Scourfield, P. (2007) 'Are there Reasons to be Worried about the "Caretelization" of Residential Care?' *Critical Social Policy*, vol. 17, no. 2, pp. 155–80.

Secretary of State for Health (2000a) *The NHS Plan: A Plan for Investment; A Plan for Reform* (London: The Stationery Office).

—— (2000b) *The NHS Plan: The Government's Response to the Royal Commission on Long Term Care* (London: The Stationery Office).

Secretary of State for Wales (1998) *Better Health – Better Wales* (Cardiff: The Welsh Office).

Seebohm Report (1968) *Report of the Committee on Local Authority and Allied Personal Social Services* (London: HMSO).

Shanas, E., Townsend, P., Wedderburn, D., Friis, H., Milhoj, P. and Stehouwer, J. (1968) *Old People in Three Industrial Societies* (London/New York: Routledge and Atherton).

Sheldon, J. H. (1948) *The Social Medicine of Old Age* (Oxford: Oxford University Press).

Sherrard, M. (1978) *Report of the Committee of Inquiry into Normansfield Hospital* (London: HMSO).

Silva, E. (2007) 'What's [Yet] to be Seen? Re-Using Qualitative Data', *Sociological Research Online*, vol. 12, issue 3, http://www.socresonline.org.uk/12/3/4.html.

Sinclair, I. (1988) 'Residential Care for Elderly People' in I. Sinclair (ed.) *Residential Care: The Research Reviewed* (London: HMSO).

Sinclair, I., Parker, R., Leat, D. and Williams, J. (1990) *A Kaleidoscope of Care: A Review of Research on Welfare Provision for Elderly People* (London: HMSO).

Social Research Association (2003) *Ethical Guidelines* (London: SRA).

Soule, A., Babb, P., Evandrou, M., Balchin, S. and Zealey, L. (eds) (2005) *Focus on Older People*, Office for National Statistics (Basingstoke: Palgrave Macmillan).

Stacey, M. (1960) *Tradition and Change: A Study of Banbury* (Oxford: Oxford University Press).

Stacey, M., Batstone, E., Bell, C. and Murcott, M. (1975) *Power, Persistence and Change: A Second Study of Banbury* (London: Routledge and Kegan Paul).

Summerfield, P. (1998) *Reconstructing Women's Wartime Lives* (Manchester: Manchester University Press).

Sumner, G. and Smith, R. (1969) *Planning Local Authority Services for the Elderly* (London: Allen and Unwin).

Sutherland Report (1999a) *With Respect to Old Age: Long Term Care: Rights and Responsibilities* (London: The Stationery Office).

—— (1999b) *With Respect to Old Age: Long Term Care: Rights and Responsibilities*, Research Volumes 1–3 (London: The Stationery Office).

Sutherland, S., Clark, J., Goodison, N., Heath, I., Marshall, M., Rayner, C., Ridley, P., Stout, R. and Wendt, R. (Royal Commissioners) (2003) *Long–Term Care*, http://image.guardian.co.uk/sys-files/Society/documents/2003/09/29/Royal-commissionersstatement.do.

Sweetman, P. (2009) 'Towards an Ethics of Recognition? Issues of Ethics and Anonymity in Visual Research', paper presented at the 1st International Visual Methods Conference, University of Leeds, 15–17 September.

Thane, P. (2000) *Old Age in English History: Past Experiences, Present Issues* (Oxford: Oxford University Press).

—— (2005) *The Long History of Old Age* (London: Thames and Hudson).

Thompson, P. (1975) *The Edwardians: The Remaking of British Society* (London: Routledge).

Thompson, P. and Townsend, P. (2004) 'Reflections on Becoming a Researcher: Peter Townsend Interviewed by Paul Thompson', *International Journal of Social Research Methodology*, vol. 7, no. 1, pp. 87–97.

Tinker, A., Wright, F. and Zeilig, H. (1995) *Difficult to Let Sheltered Housing* (London: HMSO).

Titmuss, R. (1958) 'Pension Systems and Population Change' in R. Titmuss, *Essays on the Welfare State* (London: Unwin University Books).

Townsend, P. (1957) *The Family Life of Old People* (London: Routledge and Kegan Paul).

—— (1962) *The Last Refuge: A Survey of Residential Institutions and Homes for the Aged in England and Wales* (London: Routledge and Kegan Paul).

—— (1973) *The Social Minority* (London: Allen Lane).

—— (1979) *Poverty in the United Kingdom: A Survey of Household Resources and Standards of Living* (London: Penguin Books and Allen Lane).

—— (1981) 'The Structured Dependency of the Elderly', *Ageing and Society*, vol. 1, no. 1, pp. 5–28.

—— (1993) *The International Analysis of Poverty* (Hemel Hempstead: Harvester Wheatsheaf).

—— (2005) Sheena Rolph, Julia Johnson and Randall Smith in conversation with Peter Townsend at the annual conference of the British Society of Gerontology, University of Keele, 14 July.

—— (2007) Personal Communication, 15 July.

Tryhorn, C. (2008) 'On the Mend', *The Guardian*, 30 August, p. 40.

Turner, M. and Beresford, P. (2005) *User Controlled Research: Its Meanings and Potential* (Shaping Our Lives and Centre for Citizen Participation: Brunel University).

Twigg, J. (2006) *The Body in Health and Social Care* (Basingstoke: Palgrave Macmillan).

Valios, N. (2009) 'Creature Comforts', *Community Care*, issue 1754, 22 January, pp. 26–7.

Vallelly, S., Evans, S., Fear, T. and Means R. (2006) *Opening Doors to Independence: A Longitudinal Study Exploring the Contribution of Extra Care Housing to the Care and Support of Older People with Dementia* (London: Housing Corporation and Housing 21).

Victor, C. (2005) *The Social Context of Ageing: A Textbook of Gerontology* (London: Routledge).

Victor, C., Scrambler, S., Bowling, A. and Bond, J. (2005) 'The Prevalence of, and Risk Factors for, Loneliness in Later Life: A Survey of Older People in Great Britain', *Ageing & Society*, vol. 25, part 3, pp. 357–76.

Victor, C., Scrambler, S. and Bond, J. (2009) *The Social World of Older People: Understanding Loneliness and Isolation in Later Life* (Maidenhead: Open University Press).

Wadsworth, M. (2002) 'Doing Longitudinal Research' in A. Jamieson and C. Victor (eds) *Researching Ageing and Later Life* (Buckingham: Open University Press).

WAG (Welsh Assembly Government) (2006) *The National Service Framework for Older People in Wales* (Cardiff: WAG).

Wagner, G. (1988) *Residential Care: A Positive Choice* (London: HMSO).

Walker, A. (1981) 'Towards a Political Economy of Old Age', *Ageing and Society*, vol. 1, part 1, pp. 73–94.

—— (1985) *The Care Gap: How Can Local Authorities Meet the Needs of the Elderly?* (London: Local Government Information Unit).

—— (2007) 'Why Involve Older People in Research?' *Age and Ageing*, vol. 36, no. 5, pp. 481–3.

—— (2009) 'Aging and Social Policy: Theorizing the Social' (pp. 595–613) in V. Bengtson, D. Gans, N. Putney and M. Silverstein (eds) *Handbook of Theories of Aging* (second edition) (New York: Springer).

Walmsley, J. (1998) 'Life History Interviews with People with Learning Disabilities' in R. Perks and A. Thomson (eds) *The Oral History Reader* (London: Routledge).

Wanless Report (2002) *Securing Our Future Health: Taking a Long-Term View* (London: HM Treasury).

—— (2006) *Securing Good Care for Older People: Taking a Long-Term View* (London: King's Fund).

Watkins, T. (1971) *Report of the Farleigh Hospital Committee Inquiry* (London: HMSO).

Way, A. (1984) 'A Study of 5 Residential Homes for the Elderly in Norfolk' (University of East Anglia: Unpublished dissertation).

Weaver, T., Willcocks, D. and Kellaher, L. (1985) *The Business of Care: A Study of Private Residential Homes for Old People*, Research Report No. 1 (London: CESSA, Polytechnic of North London).

Wilkin, D. and Hughes, B. (1980) *Residential Care of the Elderly: A Review of the Literature*, Research Report No. 2 (Manchester: Departments of Psychiatry and Community Medicine, University of Manchester).

—— (1987) 'Residential Care of Elderly People: The Consumers' Views', *Ageing and Society*, vol. 7, part 2, pp. 175–201.

Wilkin, D. and Thompson, C. (1989) *Users' Guide to Dependency Measures for Elderly People* (Sheffield: Joint Unit for Social Services Research, University of Sheffield).

Willcocks, D., Peace, S. and Kellaher, L. (1987) *Private Lives in Public Places* (London: Tavistock Publications).

Williams, G. (1967) *Caring for People: Staffing Residential Homes* (London: Allen and Unwin).

Willmott, M. and Young, P. (1960) *Family and Class in a London Suburb* (London: Routledge and Kegan Paul).

Withnall, A. (2006) 'Exploring Influences on Later Life Learning', *International Journal of Lifelong Education*, vol. 1, no. 1, pp. 29–49.

Wright, F. (2003) 'Discrimination against Self-Funding Residents in Long-Term Residential Care in England', *Ageing & Society*, vol. 23, part 5, pp. 603–24.

Wright, F., Tinker, A., Harron, J., Wojgani, H. and Mayaguitia, R. (2009) 'Some Social Consequences of Remodelling English Sheltered Housing and Care Homes to "Extra Care"', *Ageing & Society*, vol. 29, part 1, pp. 135–53.

Wyvern Partnership and University of Birmingham (1977) *The Provision of Residential Accommodation for the Elderly Mentally Infirm*, Research Report (Birmingham: University of Birmingham and Wyvern Partnership).

Young, M. and Willmott, P. (1957) *Family and Kinship in East London* (London: Routledge and Kegan Paul).

Yow, V. R. (1994) *Recording Oral History: A Practical Guide for Social Scientists* (Sage: London).

Index

Bold page numbers indicate photographs.